AUSTRALIANS AT WAR
IN THE AIR

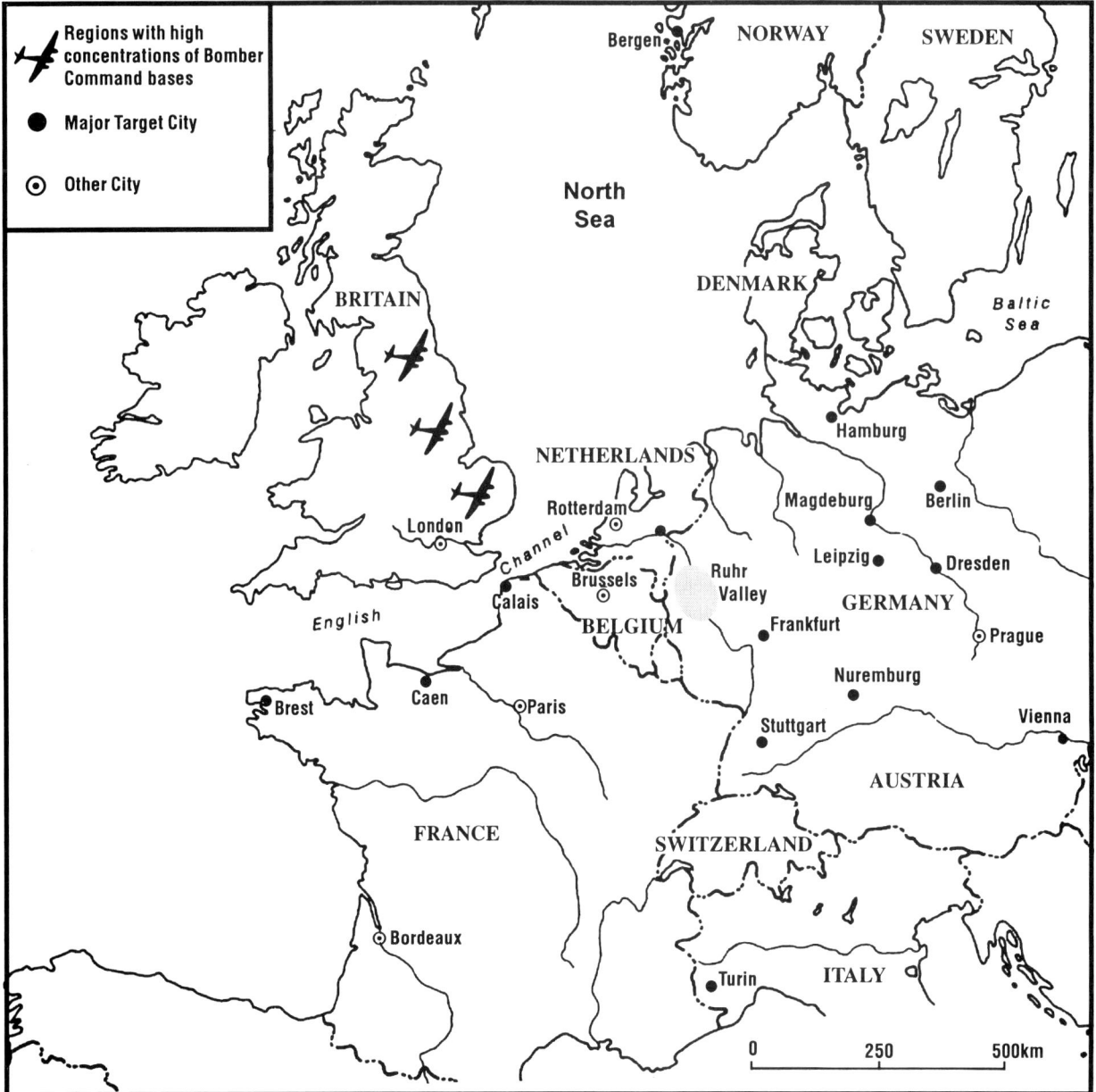

Regions with high concentrations of Bomber Command bases

● Major Target City

◉ Other City

NORWAY

SWEDEN

North
Sea

Bergen

DENMARK

Baltic
Sea

BRITAIN

Hamburg

NETHERLANDS

Magdeburg

Berlin

Rotterdam

Leipzig

Dresden

London

Ruhr
Valley

GERMANY

Channel

Brussels

Prague

Calais

BELGIUM

Frankfurt

English

Nuremburg

Caen

Paris

Vienna

Brest

Stuttgart

AUSTRIA

FRANCE

SWITZERLAND

Bordeaux

Turin

ITALY

0   250   500km

AUSTRALIANS AT WAR IN THE AIR

1939 -1945

VOLUME ONE

• GOING TO WAR • TRAINING • BOMBER COMMAND
• PRISONERS OF WAR

ROSS A. PEARSON

Kangaroo Press

Author, Ross Pearson, in uniform as sergeant
wireless air gunner.

Cover art: The Rear Gunner by Dennis Adams

Cover design by Darian Causby

First published in 1995 by Kangaroo Press Pty Ltd
3 Whitehall Road Kenthurst NSW 2156 Australia
P.O. Box 6125 Dural Delivery Centre NSW 2158
Printed in Hong Kong through Colorcraft Ltd.

ISBN 0 86417 708 9

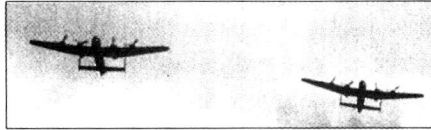

C O N T E N T S

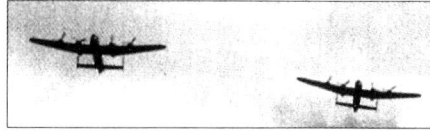

INTRODUCTION

As a member of the RAAF recruited under the Empire Air Training Scheme I served as a wireless air gunner with the RAF in Europe during 1943-1945.

I was not attached to an Australian squadron during the period. The largest number of Australians serving on the squadron (102 RAF) at any one time while I was there did not exceed twenty. Hence, when the war was over, there was little chance of a meaningful reunion with ex-members of the squadron in Australia. There were, of course, reunions in England.

With the formation of the Halifax Branch of the Air Force Association I began, from 1989, to attend reunions and was struck by the number of amazing experiences recounted at these gatherings by ordinary blokes. I started to tape some of these and gradually became more and more involved. I widened my story gathering to include Fleet Air Arm types, ex ground staff, Coastal Command, Transport Command, Tactical Air force and Middle East ex-servicemen. In keeping with my aim to be representative of all spheres I also sought out those who served in Australia, the South-West Pacific and South-East Asia.

So my stories grew in number: some oral history, some diary excerpts, some written reminiscences; but all from primary sources — Australian ex-servicemen or ex Royal Air Force men now resident in Australia.

There has been little editing of the original versions, and then only to fit the pieces into the larger picture. Some explanatory comments have been added, within square brackets.

As a frame, I have added detailed introductory sections which will provide more background to these tales. This does not, however, pretend to be a comprehensive history, but rather a series of recollections and experiences which tell of the hardships and dangers faced, and the reactions, of ordinary blokes caught up in a war.

Because of the wealth of material, it has been necessary to publish in two volumes. This volume deals primarily with the war in Europe, although there are chapters on training in Australia, on the process of going to war, and on prisoners-of-war. (The next volume will deal with the Coastal, Middle East, South-east Asia, South-West Pacific (which includes Australia) Commands and with the Second Tactical Air Force. A chapter is devoted exclusively to Ground Staff activities.)

This book would never have proceeded beyond the germ of an idea without a great deal of help.

Encouragement to proceed was given by my wife, Dorothy, and by Christine Mansell, the then editor-in-residence, Royal Australian Historical Society. My friend Jack Stronach kept urging me on when I felt the task too daunting.

The heavy job of providing linking material, presenting the stories in an interesting format and correcting the many errors of grammar and logic was made easy by the criticism, revision, redirection and rearrangement suggested by my brother, K.W. Pearson, AO, former Director of the Australian War Memorial.

The painful work of typing the original format was laboriously undertaken by my wife. The layout, committing to computer files and technical editing was completed by my friend of sixty years, Alan Stutter, DFC, whose comments were much appreciated, and most often adopted.

I would like to gratefully acknowledge the permission given by the Australian War Memorial to reproduce details of aircraft performance, armament and crewing from the publication *Air War Against Germany and Italy, 1939-1943*. Permission to reproduce material from the Murdoch Sound Archives, namely Geoff Coombe's narrative, is also appreciated.

Ross A. Pearson
January 1995

1

GOING TO WAR

Airmen trained or fought in many parts of the globe. Particularly in the early years of the war, simply getting to these destinations could be extremely dangerous. Whether dangerous or not, however, the journeys were by no means Cook's tours although there is no doubt that many were exciting and, if you were lucky enough, enjoyable.

TRAVEL FROM AUSTRALIA TO THE USA, CANADA AND THE UK

The Empire Air Training Scheme

Under an agreement involving the UK, Canada, Australia and New Zealand, in late 1939 it was agreed, that Australia would supply 10 000 aircrew every year. Of these 200 were to go every month to Canada for advanced training. The final destination for the majority was to be the UK. But wherever the outcome, travel at that time could be interesting!

Jack Dempsey (in vest) spars with Graham Etherington as he entertains Australian troops en route to Canada.

Andy Emmerson sailed from Sydney:

We went over to America on the *Matsonia* — one of the American prewar cruise ships. When we went aboard we were given hammocks down in the lower decks. They wanted people with orderly room experience to help, so a mate and I volunteered. It got us out of kitchen duties and other fatigues — although we had to take our turn up on the gun mountings.

One day we found an empty cabin. We managed to get the window open and got in. It was sheer luxury. Even the toilets had electric heaters in them.

No one found out about it until we got to Hawaii. Then they seemed to admire our initiative and left us there.

Peter Balderston also sailed from Sydney:

The Japs were in the war when we sailed. We embarked in a ship called the *Mt Vernon* — an American boat — to Wellington, then straight to San Francisco. There were four to a cabin — comparatively comfortable. The food was good. Before we left Sydney Harbour they opened an ice-cream bar and you could buy a container of ice cream — about a litre — for five cents US — all flavours. I remember I bought one and couldn't finish it so I chucked it overboard.

In San Francisco, Slapsie Maxie Rosenbloom, who had once been in Australia, owned a nightclub. He invited all of us to attend for a free night on the night we arrived. We went there and it was the first time I had drunk rum. He had a concoction called a Black Zombie which was in a tall glass, thin at the bottom and opening out at the top. The drink was solid rum — different types in layers with the thickest at the bottom and getting lighter as it got to the top. The top layer was delicious. It went down like honey. Each customer was only allowed one per night — two would blast your head off.

That same night we were put on the train and went off over the Rockies and across America — the first time I'd seen snow. We went through Salt Lake City, Denver, Omaha, Lincoln, Chicago to New York. We were billeted at Fort Hamilton, an army base in Brooklyn. We used to get the train into the city each day.

Unfortunately the officer in charge of us from Sydney had to leave us at San Francisco and he paid us all our pay to keep us going. It was about three months pay and we had a week and a bit in New York. We blew the lot in the first week by staying at the Waldorf Astoria.

We went to England on the *Mataroa*. It was a cargo ship of about 8000 tons. We were in cabins but down in the hold there were British Guianeans who were negroes. It was a refrigerated ship and all they had done was put planks over the oranges or whatever else they were carrying. The black people were a Forestry unit going to Northern Ireland. They slept in hammocks over the oranges.

We played five hundred in our cabin from morning to night and I've never played a game of five hundred since.

There must have been fifty ships in our convoy and we lost thirteen on the way over. It took us seventeen days to cross the Atlantic. The route went right up practically to Iceland and Greenland and it was mid-winter — November/December. We had P/T up on deck every morning. We were so cold we wore just about all our clothes. I had so much on I couldn't do up my double-breasted overcoat — I had to tie it up with string.

We were given icepicks and used to go round chipping ice off the superstructure. We were also drafted as lookouts for aeroplanes but our aircraft knowledge was primitive and we wouldn't have known a hostile aircraft if we'd seen one.

Ted Priest also travelled on this ship and he recalls a 'strike' on board:
We sat down to the midday meal. It amounted to one tin of peaches to every three people. We went on strike. They then served a more respectable meal and we were served better meals for the rest of the trip.

Ted Eagleton sailed for England in the MV Tolten:
It was 1941 and the Americans had come into the War. They were escorting us — only eighteen airmen on board our ship which was very small — 6000 to 8000 tons.

On the way out, after two nights, the engine broke down. We lost the convoy and went into Newfoundland. We had some repairs to the engine and hobbled back to Halifax.

As we were going into Halifax another convoy was coming out so we joined it and off we sailed again.

About four nights out we ran into a snowstorm and fog. In the night time it was absolutely terrible — the ships' sirens were going. It seemed like fifty ships in the convoy.

We happened to be on deck and saw another ship approaching and we collided. We had our port bow stove in and a few lifeboats were smashed so again we dropped out of the convoy. With all this hullabaloo the engine was damaged so we decided to go back to Halifax again.

This time we were taken off the ship and after four to eight hours we put into a larger ship — a 10 000 tonner called the *Laetitia* — and on this occasion we

Queen Mary as a troopship.

got across to England with no problems at all. We didn't arrive there until Christmas Day, which we were upset about. We finished up having Christmas dinner on the ship in Liverpool Harbour.

Gerry Judd:
We travelled right across Canada from Jasper to Halifax. We were supposed to go to New York and on to the *Queen Mary*. But this was when the *Normandie* was sabotaged in New York and capsized. This threw all the shipping arrangements in New York out of kilter — this was about early 1942.

To add to the problem, this was just when the *Queen Mary* did a zig and a light cruiser did a zag and was cut in half by the *Queen Mary*. The *Mary* then had to go into New York and be checked for damage to the front of the ship. These disasters held us up. We were stationed in a little American camp just outside Worcester in Massachusetts, which is a really pretty place. Everyone wined and dined us. Then we boarded the *Queen Mary*.

There were about 14 000 American soldiers on board and about 250 airmen. They gave you tickets for your meals and you had to be there dead on time. If you didn't get there you missed out. It was a shuttle service which began at 6 a.m. and didn't finish until nine o'clock at night. You had to have your blue, red or yellow ticket and be in the queue at the nominated time.

A friend of mine, Harry Turner, found out there was a dumb waiter going from one deck down into the dining room which used to be the first class dining area. The dumb waiter was just big enough to take one person if he folded himself up. So Harry said 'For God's sake don't tell anyone else.' So we'd go along any old time , Harry would press the button, the dumb waiter would come up and I'd get in and go down and Harry would do likewise. We did this the whole way and avoided the queues.

We travelled across to Greenock in Scotland and then down to Bournemouth. There we were staged in some old homes which had been boarding houses previously. The sergeants went to places like the Durley Dean [a popular boarding house before the war].

We arrived there on the Thursday and on the Sunday had a rude awakening to the war. Some FW 190s came over. They knew the place was crammed with aircrew and they unloaded bombs onto Bournemouth on the Sunday morning.

I remember because I was on air-raid warden duty on the roof of the Durley Dean. When the guns opened up I didn't know what the planes were. I saw them streaking across the sky and suddenly recognised them and shouted 'Air-raid! Air-raid!'.

All the blokes heard the guns going off and started to run down the stairs. Being an air-raid warden I stood at the bottom of the stairs and said 'Get your tin hats'. They had to go back up stairs and get their helmets.

They suddenly said, 'We could have been bloody killed running up those stairs.'

It was the main commercial centre of Bournemouth which was bombed out. A department store was set on fire. It was a Sunday morning and all the hotels used to open for a while. Quite a few airmen and soldiers as well as some of our aircrew had gone into a hotel which was on the corner.

On the higher side of the street was quite a large church with a large tower. A bomb hit the spire of the church and was deflected down into the hotel and killed all these fellows. When you weren't prepared for this sort of thing it came as a shock.

Fred Wright:
I sailed on MV *President Washington*. There were six to a cabin so it was relatively comfortable. During the trip aircrew were given different jobs to do on the ship, like keeping a watch out. I was delegated to KP duties just as we were crossing the Equator.

We were sent down to get turkeys out of the meat room and when we got there we found three bodies. They were American sailors killed somewhere in the East being taken back to America for burial. We went from Los Angeles to the East by train — Camp Miles Standish on the East coast. At this stage seven went down with mumps and we were kept there for another two weeks. During this period we were billeted out with American families.

Then we embarked on the *Aquitania* which went across the Atlantic on its own because of its speed. During this trip, before we turned north-west, we were given duties as aircraft spotters.

As we got almost to the Bay of Biscay to go round Ireland into Greenock the boys on spotting saw a four-engined aircraft come across. They didn't bother to raise the alarm as they thought it was a friendly aircraft. The guns started banging off — it was a Focke Wulf Kurier which they had not recognised. He came in, fired his guns and turned away.

Alby Silverstone followed a similar route to Fred Wright:
Prior to the entry of the Japanese into the war and the consequent involvement of the USA, the Americans continued their normal shipping schedules to the countries of the world with whom they traded. This included fortnightly sailings of the *Mariposa* and *Monterey* between Sydney and San Francisco.

In that time, Australian airmen bound for Canada had been transported as far as San Francisco on these ships as normal passengers and then on to the various camps by normal transport. It had also been usual for these airmen to be accorded tourist treatment while in transit — things like visits to film studios and the city sights. All of this changed with the new situation brought on by the Pacific War.

The new method of reaching Canada was by taking the return trip of the various ships which brought troops to Australia.

The *Tasker H Bliss* had originally been named the *President Cleveland* and had then been a 10 000 ton passenger vessel. It was said to be of 1921 vintage and was fairly aged as passenger ships go. It was now considerably changed to meet troopship conditions.

There were no longer cabins available for Australian airmen — only troop arrangements. Despite the fact that our 200 airmen and a handful of US troops (mostly wounded) were the only servicemen on board the ship, the cabins were left empty. Initially we were deposited in four-tiered wire bunks situated in the forward hold and accessed by twenty-foot steel ladders. Later, as we went through the tropics, the heat of the holds became unbearable and we were allowed to move to deck areas.

After leaving Sandgate on 16 June we sailed out into Moreton Bay and were escorted into the Pacific for the first 200 miles by a destroyer. After that we were on our own.

The first few days went smoothly except for the cramped sleeping conditions. But, as the bunks were only for sleeping, the conditions were somewhat tempered by the fact that we had the complete run of the ship — particularly as there were less than 300 troops on board compared with some 5000 on the outward trip.

About five days out I reported to sick bay with a heavy cold. It was diagnosed as a slight fever and I was hospitalised for five days. Towards the end of this stay the ship ran into a violent Pacific storm. It raged for about three days during which the ship tossed and pitched rather badly. On the last day of the storm the nursing sister decided I should have a dose of Epsom salts. It went down and came straight up again and that started my stomach on a bout of seasickness.

I was discharged from Sick Bay that same afternoon and then went through the worst week of my life. The storm slowly subsided and we ran into calmer waters but my seasickness continued. It was pure agony and for days I just laid in a heap. I didn't care if the ship floated or sank. I couldn't eat and even the smell of food sent me hurrying to the ship's rail.

For five days I existed on dry biscuits from the ship's PX store (canteen) and after a week of nice calm weather was at last able to look at food.

While we had no compass or navigation instruments, it was obvious from the position of the sun that we had headed far to the south. Over the next few weeks all the trainee navigators passed part of each day estimating our position and distance which we checked against a huge map of the world that hung in the ship's lounge.

The days now passed pleasantly and, after the earlier storm, the weather behaved perfectly. We eventually moved away from our south east course and moved to a northerly direction. As the weather became warmer and we approached the tropics we were moved from the hold to bunks on the promenade deck. Special days were celebrated — such as the 4th of July (American Independence Day) and a special ceremony was laid on for 'crossing the line' complete with King Neptune and his retainers. I even received a certificate to commemorate this, per medium of the ship's Gestetner.

Church was held each Sunday, conducted by the ship's chaplain, and many of the ship's officers gave talks on topical subjects. Amongst the few Americans returning to the United States was Joseph Harsch, foreign correspondent for *The Christian Science Monitor*, who had been making a tour of the Pacific war zone. He kept us up to date on the war news as received by the ship's radio. We even had a concert or two organised by our own people who had formed an entertainment group.

We were into our fourth week before we saw sign of land, and then only obliquely, with the sighting of birds, porpoises and flying fish. It then became obvious from our direction that we were heading for a port and we guessed it might be Panama. Firstly we came to the twin cities of Balboa and Panama City and then passed through the canal and into Christobal Harbour. There we docked for several days while the ship awaited further orders.

Christobal was both the entrance to the Panama Canal and the gateway to the Caribbean Sea, depending on which way you were travelling. At this time the Caribbean was the most dangerous area for shipping in all the war zones. In the previous six months more than 180 allied ships had been sunk there and it was fairly obvious that we would need to wait for the formation of a convoy. It was at this time that we learned that our presence on the *Tasker* was not known at US Naval Headquarters. On contacting Washington for further orders on our destination and where to unload the 'Australian airmen' the OIC Troops received the puzzled reply: 'What Australian airmen?'

We were allowed ashore at Panama for an afternoon tour with an American MP guide and, for this purpose, were split into small groups of about six. It was for all the world like the movies version of a tropical town, which of course it was. Even the bars had that bamboo look Hollywood used for island nightclubs.

We eventually moved into the Caribbean as part of a huge six-knot convoy with escort from destroyers and Catalina flying boats.

It took a further three weeks to reach our port of disembarkation after calling at Guatemala, Cuba, Key West (at the bottom of Florida) and, finally, Norfolk, Virginia. The journey was not without incident and several submarine attacks were turned away by our escorts although not without the loss of two small freighters. In all, our journey from Brisbane to Norfolk had taken almost eight weeks.

At Norfolk we boarded a train which transported us north to Lachine, Montreal, Canada and, on the way, confused all sorts of people who had never heard of Australia, let alone met any Australians.

We later found that the *Tasker* had been switched to the Atlantic as part of planned North African landings. It was sunk during that campaign.

Ross Pearson's trip to England was none too pleasant either. Let him tell the story:

We sailed from Port Adelaide — much to our surprise. We had entrained in Sydney, spent three days in Melbourne and then travelled on to Adelaide. We presumed we were bound for Darwin.

Our surprise was heightened by the nature and type of ship. She was a small vessel with limited passenger accommodation and could barely be seen above the wharf at Port Adelaide. The *Denbighshire*, a Dutch vessel, had carried a load of meat in the hold which was now to accommodate a cargo of airmen. The hold smelt of the former cargo. It was heavily blacked-out - which prevented air getting to those travelling below.

We sailed out from Port Adelaide and headed south. I was to create an unenviable record. From Australia to the United Kingdom I was continually seasick. While others prayed for Deliverance, I prayed for the ship to sink to put me out of my suffering. Each day I took my place at the rail and each day I dutifully delivered what little I had eaten to the deep. I am sure that all that kept me alive was the oranges provided by my friends who gave their ration to me.

David Corthorn:

We came across Canada by train. When we got to Halifax we waited for a ship. In our holding camp we found we had to salute the Canadian flag. A lot of the Australians would not do this. The Canadian CO was annoyed about this so he used to get us up at six o'clock in the morning and we would march out to salute the flag. It wasn't quite 60 below but 30 to 40 — or so it seemed to us.

For our sins he confined us to barracks but we pulled down a section of the chainwire fence and visited our friends in Halifax.

All the buildings were timber and were well heated. In case of fire strong axes were provided to cut a way out. One of our chaps — we called him 'Dorrigo' because he came from there — was a timber-getter. He looked at one of the axes and said, 'How about we chop down the flagpole? Then we won't have to salute the flag.' We thought he was having us on.

So he said, 'If you're game and come out with me, I'll chop it down.'

He did!

We went out very early one morning. It only took him about four chops — he did a beautiful job. It just fell slowly and the back end of it just fell down but not on to the ground. It just sat there. At 6 a.m. we all went out to salute the flag, with the CO coming personally.

It wasn't very light at that hour and the CO didn't realise what had happened to his flagpole. An NCO had the flag draped over his arm and said, 'Sir, I can't put it up.' 'Why not?' asked the CO. He turned round and nearly blew up. 'Who did this?' No answer. So he said, 'You're all confined to barracks!' Someone said, 'Wrong again.'

Next morning he had decided to give us a route march out of Halifax and back. So we got ready in full uniform — mostly all NCO's. All the French drill instructors were French Canadians. It was rumoured they wouldn't allow them in the fighting services because they didn't trust them — they were all right to drill you.

They got us out on the parade ground and called our names and marched us to the saluting area.

Some fellows had a bright idea — they would act like sheep. They started to 'baa' and bark and whistle. We finally got to the saluting area sounding like a mob of sheep. When we reached there the cacophony was tremendous. Then silence. Confined to barracks AGAIN.

But the *Louis Pasteur* came in and we were marched down to the ship. It had had German POWs on board and it was filthy. It was lousy with cockroaches so we got off again. The CO said, 'You've got to get on!' but we said, 'No! Not until it's fumigated!'

Well, this caused a real upset. They cabled Australia and got a senior officer over. They flew him over in a Liberator — it must have taken three days to get to Halifax. He landed and wanted to know what it was all about. He went on board and said, 'I wouldn't put pigs on it!' so he told us to go back to sleeping quarters until it was fumigated.

Harry Brabin tells of his journey from the States to England:
The ship was the *QE*, capable of sailing to England in three days. It was jampacked with troops so no exercise on deck was possible. Crap, crown and anchor, poker and other gambling games were being played everywhere.

We had only two meals each day. The food was placed at the head of the tables and those closest served themselves and passed the dishes on, so that those sitting near the bulkheads were sometimes left hungry. 'Chow hounds' was the name given to soldiers who used extra plates and pannikins.

Four of us slept in four-decker bunks in a storeroom below the waterline. There was nowhere to put kitbags except at the foot of each bunk. For the first time in my life I was glad to be short as the tall men really had great difficulty sleeping.

AUSTRALIA TO THE MIDDLE EAST AND AFRICA

Not all aircrew went to the UK. Some went by boat to Africa. As Ron McCathie notes:
Part of an Air Force contingent, together with 5000 Army reinforcements for the seventh division in Egypt, I found myself aboard the *Queen Elizabeth* in Athol Bight (Port Jackson).

We sailed from Sydney in convoy with the *Queen Mary* and the heavy cruiser *Cornwall* as our escort. During the day the 'Queens' sailed abreast with the *Cornwall* leading. At night the configuration changed to line astern, always with our escort up front.

Crossing the Great Australian Bight we ran into a storm with the wind from the south-west and waves up to thirty-five feet.

Our speed was twenty-six knots. The ships displaced 80 000 tons and the sight of the *Mary* ploughing into this seaway only half a mile away was spectacular. At the bow it was eighty feet to the waterline and every now and then she would take a green sea over the bow that would race back to the bridge and explode. All those tons at this speed were difficult to fully appreciate. Of course the *Elizabeth* was doing the same thing but, surprisingly, the motion aboard was really moderate. The poor old *Cornwall*, with her low freeboard, was practically underwater all the time and having difficulty maintaining station.

Six of us had been detailed for blackout duty at sundown and each day, after rounds, reported to Captain Lasenby in the Main Lounge — he bought the drinks. One of us asked why we did not slow down and 'give the *Cornwall* a fair go'? Lasenby intimated that these two

ships never stopped or slowed down, no matter what the circumstances. If you fell overboard the best you could expect was an entry in the ship's log 'Man overboard 2000 hours, Lat..., Long..., Presumed drowned.'

The food was excellent and the bars opened three times each day. Pink gin or a scotch cost fivepence.

The only universal complaint was the atmosphere once we were battened down for the night. The air-conditioning systems were never designed to cope with so many people. The odours and heat came from the turbines at full speed, cooking (which must have gone on for most of the time), together with the body odours from all of us. Add on a fart or two from the 5000 troops below and, by morning, you could almost cut the air with a knife. It became worse as we moved into the tropics. We had an epidemic of what the medics called 'blackout fever'.

The convoy stopped for twenty-four hours for fuel and water at Trincomalee, the Royal Navy's base at the southern end of Ceylon.

We arrived at our destination, Port Taufiq at the southern end of the Suez Canal, early one morning and disembarked.

The port was a shambles. Sunken ships everywhere, masts and funnels leaning drunkenly in all directions and the dockside still smouldering as the result of an overnight raid by the Luftwaffe — not very reassuring for the incoming rookies.

I understand the 'Queens' were turned round in less than twelve hours. After sundown the Ju 88s would arrive to continue their nightly bombing raids.

Troop trains took us overnight to Alexandria where the RAAF contingent spent ten days billeted at the Hotel Victoria, Abu Kahir — it smelled like a camel. We spent most of the time enjoying ourselves in Alexandria.

With Rommel descending on Cairo — as he thought — RAF Headquarters in Cairo decided to return us to Port Taufiq. We found ourselves aboard the peacetime Indian Army troopship *Dilwarra*—a pretty little ship of about 5000 tons and beautifully appointed.

In a short time we arrived in Mombasa after picking up Ethiopian Italian POWs at Port Sudan en route. They were lucky to finish up in Kenya. They spent the rest of the war building roads — and very good ones too.

Christmas Day 1941 we spent sailing down the east coast of Africa. The ship's victualling and cooking departments provided the best of food and liquid refreshments. It was a very happy day. Even each of the POWs enjoyed an extra packet of cigarettes and a bottle of beer.

Mombasa streets and gardens were a mass of frangipanni trees. They were all in bloom and the perfume delightfully scented the balmy evenings. The fireflies were so numerous they almost lit up the streets. I never catch that lovely perfume without memories of that lovely small East African port.

SOUTH AFRICA TO BRITAIN

In October 1940 Australia agreed to send a draft of forty airmen to train as pilots in Rhodesia. A dangerous journey by ship to the UK faced them on the completion of their training.

The two following stories link together quite by accident. The first is an extract from the diary of John May (later killed in action) who tells of boarding the **Nea Hellas,** *an old one-funnelled 17 000 tonner, bound for England:*

9/10/42 To-day has been a day of great panic. Last night, over the radio, they heard that five boats had been sunk, one only seventy miles ahead of us, including the *Oronsay* which was next to us at Cape Town with 6000 troops aboard. . . The Captain spoke over the wireless to all. 'We're in very dangerous waters. . . all to sleep fully clothes. . . have a "panic" bag packed. . . carry lifebelt . . . keep sharp lookout.'

The Germans are using Dakar, 200 miles away, as a base and are watching Freetown like terriers. . . crew all frightened. . . stokers up inspecting smoke. . . Between eight and nine to-night we're supposed to be looking out for survivors from *Oronsay* — 6000 troops on leave from Middle East.

Three torpedoes fired at us. One went past stern and blew up some distance off — two more were time torpedoes and blew up too soon. Close enough to shake the ship! We were lucky!

11/10/42. We're the only ship out of nine from Cape Town to reach Freetown!

12/10/42. Survivors keep coming aboard in those invasion barges — women and kids, some toddlers — and people from Argentina coming to join us in the cabins. A diver has discovered dents in the hull.

17/10. Saw barge pull alongside with survivors. . . five more boatloads still adrift — eight days in open boat.

Geoff Coombes picks up the story:
I was one of a draft who went to Alexandria on the *Queen Elizabeth* with the AIF. After a short time at Alexandria we caught a ship to Durban and then a train to Bulawayo in Rhodesia.

Rhodesia was very colonial at that time. We were 10/- a day tourists and thoroughly enjoying it when off duty.

After eight and a half hours on Tiger Moths I soloed at EFTS to train on Oxfords where I spent six months

and 120 hours before winning my wings. Life in Rhodesia was good. We were made sergeants (unpaid) to enable us to go to the sergeants' mess. The station barracks were comfortable and were staffed by 'house boys' of the Masai tribe.

As only pilots were trained in Rhodesia everyone was keen to succeed as, if anyone was scrubbed, he was returned to Australia for remustering.

On completion of training, after a short leave, the course left by train for Cape Town where we embarked on the *Oronsay* en route to the UK. Being a fast ship we were not in convoy. When sailing due north from Cape Town the weather was very hot so some of us went on deck to sleep and, as it happened, this probably saved our lives. After seven days out the ship was torpedoed at dawn. We were slammed by two torpedoes which put an end to our little cruise. They couldn't get the lifeboats down on one side. Luckily for me it wasn't where my boat station was and we finally evacuated the ship.

I understand about 200 guys were drowned. They didn't get out of the engine rooms or were downstairs. It didn't go down very quickly but the two torpedoes put the whole explosion into the engine rooms. She listed badly. Not long after we got our lifeboat down they slammed two more torpedoes into her and that made her go down fast.

I didn't see the submarine. Some of the other fellows did in other boats. They said she asked what ship it was.

We had two groups, each of six lifeboats, and each group towed by a motor launch. The weather was beautiful. At that stage there was hardly a ripple.

We thought we were about halfway to South America — about 640 miles from the African coast, and that would have been the Gold Coast and Sierra Leone. Our launch started off for the African coast. The other had a bit of trouble starting and we were a couple of miles away when our launch broke down. Then the survivors on the other launch started their engine and their group passed us and disappeared over the horizon. They got picked up after a couple of days but their group had split up and some didn't get picked up until the eighth, tenth and twelfth days.

We kept together till the morning of the eighth day and we decided we were hindering one another so we split up to make it on our own.

Our lifeboats had some crew members in them and some survivors from other torpedoed boats. There were a number of colonial English being called up and going back with their families to serve in the forces. There were not many RAAF fellows.

There were nineteen in our boat. I had transferred to this boat which initially went down with the ship and came up to the surface. So some of the rations had got wet and some of the gear was missing.

We transferred some people from other lifeboats into it and we fared rather well. Morale was good — we reckoned we would make it. We reckoned we'd rationed the water out well and we were moving in the right direction.

On the eighth day we had covered about 320 miles. Happily, we were then spotted by a Sunderland who sent a message that a destroyer would pick us up in four or five hours. They were a long five hours.

The destroyer was the *Brilliant*. The first indicator that the destroyer was coming up was the light on the horizon.

Then it would disappear and, finally, it came up firing star shells, each one getting higher until one exploded over the top of us. The shell cases seemed to hit the water near us to illuminate us.

Meanwhile we had hand-held red flares going. We were looking where the shells were coming from when suddenly a searchlight came on behind us. The destroyer had come round behind us.

They said 'Come on up the scramble nets'. We were the first boat there.

We went into Freetown and arrived there in the evening. They took us in a landing craft straight on to a ship to England. It was the *Nea Hellas* — a bloody awful old ship — survivors of fifteen ships on board.

We had an escort of destroyers all the way to England. The British Admiralty had, I think, condemned it and sold it to the Greeks who had made a sort of cruise liner out of it. Just a boat. Got to England on it anyway.

BRITAIN TO MALTA

Walter Mould tells his story:
I left Gourock on the Clyde in June 1941 on board the Irish mail (packet) steamer *Leinster* - chock-a-block with RAF and Army personnel. We proceeded in convoy out into the Atlantic, it seemed halfway to Canada, then south, then east to Gibraltar. I cannot remember how long it took, probably five or six days. I remember the unfinished *Louis Pasteur* was, by far, the largest ship in the convoy. It had been taken from Le Havre just before the fall of France. There were several merchantmen loaded with supplies for Malta. This journey was without incident — but incidents came very soon after leaving Gibraltar.

The convoy to Malta did not include the *Louis Pasteur* — she was too big and conspicuous a target. It did include an aircraft carrier, the battleship *Nelson*, several cruisers and destroyers.

After several days we set off from Algeciras Bay, confident that the German and Italian agents in Spain had informed their HQs of our departure and destination. I, as did many, slept in a reception area with lifebelt close by. It was hot below deck but we were not there for long. I was awakened by the ship coming to a sudden stop.

I thought we had either hit a mine or been torpedoed. It was soon apparent that neither had happened — there was no indication of listing or sinking. But that is not to say there was no panic or chaos. The ship was jammed fast on the rocks on the other side of the bay — we were in Spanish territory!

In the late afternoon or early evening of the day before, the ship's officers had held a birthday party for the navigator resulting in him getting us about 40 degrees off course.

We were told to collect essentials and take up boat stations. We were not allowed to bring kitbags etc. with us.

On deck, there was shouting and cursing everywhere. We could see, in the half-light of dawn, the cliffs a little way off and some bodies looking down on us. To seaward, a British destroyer was heading towards us with its guns trained on the cliffs. It came right into Spanish territorial waters. We could see Gibraltar on the other side of the bay and the straits to the south. Some thought this was a good time to hop over the side.

By chance, I was allocated to the powered lifeboat on the starboard side, that is, on the Algeciras side — the town could be seen some distance away. Our boat was to tow other boats across the bay — aided by their rowing.

The senior officer in the boat was a brigadier. I remember he kept telling us he was a VIP. The RN liaison officer was a lieutenant commander and he was cursing some of the junior officers (mostly Army captains and lieutenants with one pilot officer or flying officer) because they weren't handling the boats properly. Well, they weren't sailors.

Once in the boats, we proceeded slowly across the bay under the watchful eye of the destroyer. By then it was broad daylight and more figures appeared on top of the cliffs with what looked to be some artillery.

We arrived at Gib, were put aboard the *Louis Pasteur* and were given some breakfast.

The *Leinster* was towed off by two destroyers. I never found out what happened to her or the captain and crew. But the latter helped themselves to our belongings, as matelots are wont to do. Some of our stuff was brought to the *Louis Pasteur* where we tried to sort out what was left.

We stayed in *Louis Pasteur* for a few days and had a look round Gib, where we could buy duty-free cigs, etc. and unlimited chocolate — all severely rationed in the UK. The we were off to Malta, this time in the cruiser *Hermione*.

In the meantime the convoy had been severely mauled. At least one merchantman had been sunk together with one cruiser and one destroyer. The *Nelson* had been torpedoed. One cruiser, with dead and wounded aboard, called in at Gibraltar while we were there. I don't know all the details, but they had been attacked by Ju 88s, Italian bombers and E-boats. The E-boats were small, very fast and were armed with torpedoes.

We may have been very lucky.

On the journey to Malta we had three cruisers in line astern and were flanked by a Lightning class destroyer on one side and the Dutch destroyer *Van Tromp* (named after the famous admiral) on the other. The voyage lasted two nights and a day. We arrived off Malta early on the second day — about 10 a.m.

At dawn that morning the captain's voice had come over the intercom: 'Stand by for. . . ' Next moment the cruiser seemed to stop in its tracks — a jarring impact. It had rammed a submarine which had apparently been caught on the surface. I never found out the nationality of the sub. The RN asks the questions afterwards.

Some of us rushed on deck at the impact but could only see the coming dawn. The captain explained briefly what had happened. The British destroyer wheeled round to pick up survivors. I don't know whether the sub survived.

During our short voyage we slept where we could. If a hammock was empty we took it When the sailor came off watch we were tipped out. During the single day of the trip we walked around the deck and gazed at the North African coast around Algiers. We even walked on the quarter deck, which was normally only for officers.

We ate well — better than on a troopship or generally in the RN. Hurricanes escorted us from some way out of Malta. An attack by E-boats or Ju 88s had been expected but none turned up so we serenely sailed unmolested into Grand Harbour.

Earlier that year the harbour had been under savage attack when *Courageous* was there under repairs after receiving direct hits during a previous convoy. Carriers weren't taken that far after that.

DO IT YOURSELF

Some RAAF aircrew did not have the advantage of travel by troopship. They took themselves and their planes to their squadrons which they found in New Guinea, in the Middle East and in the Far East — to name just a few destinations.

Fred Cassidy had to take himself to war in New Guinea:
I was sitting in the tent one day when the boss said to me, 'I want you to navigate this plane to Tadji [New Guinea].' The young P/O pilot was a Beaufort pilot and he had to take a Beaufighter. He did one circuit to get accustomed to the Beaufighter — they weren't that easy to fly.

We set off and at our first port of call he started to come in at 120 mph. I thought, What's going to happen here? Anyhow, he bounced it, sank a bit and then went

Hurricane of Queen's Flight visiting Imperial War Museum, Duxford, Cambridgeshire

Beaufighter aircraft.

round again. He said, 'What speed do you bring these things in at?' By that time I knew a bit about them — my skipper used to let me fly them — no landings or take-offs though — so I talked him into a landing.

Then we went off to Garbut. Same thing there. We got up to Seven Mile at Moresby next. Same again. I thought he'd kill me.

I explained to him that the strip at Tadgi was very short with trees all round it. There would be no opportunity to go round again. So, fortunately, he did a perfect landing.

Fred seems to have made a hobby of taking people to war in New Guinea:
There was a bloke from 75 Squadron in a Kittyhawk who said to me, 'Where are you going?' I said, 'To Moresby.' He explained that he had to get the Kittyhawk to Moresby but had been ordered that he must have an escort. So I invited him to come with us. Two-thirds of the way over the weather closed in.

I called him on the R/T and said, 'You can't go back and we won't go back. So just tuck in close with us and we'll take you to Moresby.' So I got the D/F loop working and we came straight up Moresby Harbour.

Jack Stronach had to fly himself to the Middle East:
In part of April and May 1942 I was stationed at No 1 Ferry Flight at Lyneham in England testing a Hudson aircraft in preparation for flying it to India:
On 29 May I took off from Lyneham and flew to Portreath, the most south-westerly aerodrome in England. There we attended a briefing on the route we would take to India. We were to fly to Gibraltar, then down the west coast of Africa to Takoradi, across Africa to Khartoum and then north to Cairo. Just after dawn broke

on 30 May we took off for Gibraltar, flying across the Bay of Biscay in light rain and made a landfall at Cape Finisterre at the north-western point of Spain. Then we flew on down the Spanish coast to Portugal where the weather cleared. We flew just off the coast and when we reached the south of Portugal we flew into a very heavy haze and we lost sight of the coast. After a short time the haze cleared and a coastline came up. It was desert! I looked at the compass and realised we had been flying in the wrong direction. We were off the coast of Morocco and had flown past the Straits of Gibraltar in the haze.

The navigator gave me a course back to Gibraltar. On making my approach to the runway I put down the undercarriage, but when I went to put down the flaps the flap lever jammed and I was unable to lower the flaps. This meant I had to land the aircraft at a higher speed than normal. I got the wheels on the runway and started applying the brakes. The end of the runway was coming up fast and I could see the tails of other aircraft that had overshot the runway and gone into the water. Just before the end of the runway the airgunner opened the door and jumped out. Right on the end of the runway I swung the aircraft hard and stopped, looking at the water under the left wing of the aircraft.

It was not until 23 June that I was informed we were to take off at 3 p.m. that afternoon and fly down the Mediterranean Sea to a place called Mersah Matruh in Egypt.

We became airborne at 3 p.m. climbed to 10 000 feet and flew out to sea with the North African coast in sight. Darkness fell as we started to fly over Tunisia. Our course took us just south of Malta which was having an air-raid. We could see the searchlights and flak bursts.

Later our course took us over Tobruk where they fired a few shells at us.

Unbeknownst to me, a drama was taking place in the back of the aircraft. My navigator, Eric Bell, a fellow Australian, informed me at a later date that the air gunner, who came from Jamaica, had taken a bottle of whisky on board the aircraft in Gibraltar and by the time we came under fire over Tobruk the contents were in his stomach. He started to go crazy and tried to open the door and jump out without a parachute. The navigator and the wireless operator struggled with him and got him on the floor with the navigator holding him in a headlock and the wireless operator sitting on him until he was subdued.

Later, as we were approaching Mersah Matruh where I was going to make a night landing, we received a wireless message that it had been captured.

The German forces under Rommel were advancing at a fast rate into Egypt.

All I could do was fly on and hope for the best. I had been flying, at that time, for just over ten hours, all over enemy territory.

As dawn broke I started to descend, passed through a layer of cloud and came out over the sea at 2000 feet. I checked the fuel gauges and they showed empty. I turned the aircraft due south to hit the coastline of Egypt, if I could make it, and make a belly landing in the desert. Luck was with me.

As I crossed the coast a landing strip came up with Wellingtons parked along the side of it. I went in and landed, making a total flying time of thirteen hours.

The Wellington crews had just come back from bombing Benghazi and could not believe I had come all the way from Gibraltar.

I was the first aircraft to do this flight, and the last, as the Germans advanced on El Alamein.

David Corthorn faced a very long journey from England to his squadron's base in Jopore, India. As the navigator of his plane, he had to guide the pilot to the base from which they would carry out their future role.

Our first leg was to be from England to Rabat Sala in North Africa.

After we took off the weather conditions were not the best and we did not see land at all. Initially we flew at 10 000 feet. We had no oxygen so 12 000 feet was our ceiling.

En route I said to the wireless operator, 'Give me a bearing on the beacon at Scilly Isles.' He said, 'I can't do that, my radio's not working.' He'd pulled all the valves out. They were rolling round the floor like beer cans. So I worked out a DR position for the Scilly Isles and arrived there on our ETA.

But we couldn't see land anywhere so we turned roughly south going for Cape Finisterre which was the northernmost part of Occupied France, just before you come to the Bay of Biscay. [Cape Finisterre is at the north west tip of Spain, at the south of the Bay of Biscay.]

Bear in mind we only had a certain amount of petrol on board — enough to take us safely to North Africa. We had 1000 gallons in our tanks altogether. We sighted Cape Finisterre all right and I said to Charlie, my pilot, 'Do you think we should go straight across to the northernmost cape of the Iberian peninsular?' But just then we sighted what we thought was a Focke-Wulf Condor coming across the Bay of Biscay. Although we had twin Brownings in the front turret and twin Brownings in the rear turret we didn't have a mid-upper gunner and, in any case, we carried no ammunition, in case we landed in a neutral country en route.

We decided we'd head towards the Azores to keep out of German aircraft range. However I did have a Verey pistol and colours of the day in case we crossed the Navy because they'd shoot you down if you didn't fire the correct colours.

'Let's fire one of these cartridges.' I said. So I put it in — it was one or two inches in diameter — and we fired it. By now it was almost night and getting pretty dark and the colours were rather spectacular. It had the desired effect because the Focke-Wulf turned east and headed back towards northern France. We were very thankful and continued on our way.

We had gone out further west than we anticipated and I took bearings on the Pole Star because you could get a position line at right angles to our course. Our track was practically south so you could mark it off at right angles — the Pole Star being approximately due north. Just one of the things that astro navigation gives you.

So we were able to mark up our groundspeed pretty accurately even though we couldn't see land.

Our petrol gauge was showing NIL so we tilted down to see if it had anything to do with the level. But it still showed NIL so we decided to fly due east and hit the coast of Portugal. I think our airspeed was 120 knots and the wind was blowing at about 100 knots so we were doing only about 20 or 30 knots — maybe 50.

After about an hour I saw a lighthouse. Of course, all neutral lights were being used whereas those in occupied countries were not. Portugal was neutral and still using its beacons so we could identify this light and see where we were.

Dawn was breaking and the light went out so I don't know to this day where we were but I suspect it was near Lisbon. We had started at about 10 000 to 12 000 feet but when we were short of petrol we decided to keep the nose down to save petrol and maintain our airspeed even though we gradually lost height.

In addition to the crew we carried one passenger. He was a groundstaff person who was going to Casablanca in North Africa. He only looked about sixteen so I said we would have to look out for a nice beach to ditch on as we had no petrol. I shouldn't have said it because he was promptly sick in the Elsan down the back. [Elsan — A portable aircraft toilet].

Anyway, we kept on, still checking the petrol, and crossed Cape St Vincent. They had large 16-inch naval guns there and we were so low we could see them sweeping round. We were hoping they would not fire. I said to Charlie, 'Have you ever seen Gibraltar?' and he said, 'No, I've never seen it.' I said, 'I've only seen a picture postcard of it.' So we decided we'd land there.

Gibraltar was one of the most secret places on our route and you had to be directly routed there to land — if you weren't you shouldn't! It was a court-martial offence. So we decided we were running out of petrol and Gibraltar was the closest place. We had the radio operator call up Gibraltar on the HF radio and asked for permission to land. They came back and said, 'Okay. You're No 3 to land.' Then we saw the Rock and all the lovely blue water — 'Out of the blue comes the whitest wash.' We saw Gibraltar on the port side and came round. Charlie said we weren't going to be No 3 to land — 'I'll land as soon as I get the wind.'

The wind was blowing from the west so we swung around — you couldn't divert and if you overshot you would be in the water. So, down we came — flaps down and made a good landing. There were all these red lights from the control tower.

We hadn't taxied far after landing when the motors stopped and we pulled up. I said, 'What are you cutting the motors for?' Charlie said, 'I haven't cut them.'

The CO came out in his little waggon with black and white checks, put his head out and said, 'Come and see me!'

We went and had breakfast — bacon and eggs as usual — and I said to the operator who was going to refuel us, 'Tell me how much you put in!' He came back half an hour later and said, '1008 gallons.' I said, 'We only carry 1000.' So he said, 'You must have been operating on vapour.' We had used it all.

The CO said, 'Why did you land?' 'Because we ran out of petrol.' He said, 'How did you know?' We told him.

We took off on Friday and flew to North Africa and then on to Rabat, Castel Benito and Cairo.

In Cairo I thought I'd get a film for my camera. It was black and white film then but I'd heard you could get colour film. So I was looking for this place when a chap in RAAF uniform came up and said, 'Good day Rex!' I said, 'I'm not Rex, but that's my brother!' He was in 3 Squadron.

From Cairo we went to Habbaniyah in Iran. We followed the oil pipeline from Cairo to get there.

Habbaniyah was a British garrison town in the middle of Iran (then known as Persia). The Shah of Persia was pro-British.

The RAF had a big aerodrome there which was used as a staging post for aircraft coming out from England. We stayed there for four or five days. The quarters were very comfortable — carved from stone and marble with electric fans and mostly British food.

Then we went on to Bahrain. We landed on the island of Maharat where there was an airstrip.

The aircraft needed some repair and maintenance so we had a look around. We went to the Virgins Pool which was owned by the local sheik. Apparently the beautiful palace around the pool was his harem and all his wives lived there. If they didn't please him he used to toss them in the pool. The pool had carp which disposed of the unwanted bodies.

We were asked if we would like to have a trip in one of the old biplanes they used for some rescue work. A Fortress had crash landed between Bahrain and Shahjah and this chap wanted to take coffee and biscuits out to them. He was a real pukka RAF type. The crashed crew was American. When we landed all the guns had been stripped and the crew was about a quarter of a mile away. It transpired that they didn't want coffee — they had plenty of beer!

Another aircraft took them off later.

Then we were on our way to Shahjah and took off from their metal runway for Karachi. From there we went to Allahabad and on to Jopore on the India/Burma border.

TRAVEL BY LAND

Ron McCathie found himself in Persia (now Iran) at Tehran. He had to return to his headquarters overland with a ground party. At the end of six months in Tehran the Russians told D Force to leave. They had no need of bodies. All they wanted from their allies was equipment.

Consequently the unit was posted back to Egypt where we were to be disbanded.

In retrospect I think we were very lucky. Going to Russia in the middle of winter held no attraction at all.

It was, on occasion, twenty degrees below on the aerodrome with a foot or more of snow around our tents.

I elected to travel back to Cairo with the ground party in company with the captain in his fifteen-hundred-weight truck, hoping to get some shooting along the way.

The return was uneventful but for the crossing of the Shah Pass (elevation 12 000). It was the depth of winter and temperatures were sub-zero all the time. Snow drifts were often twenty feet deep.

The poor Royal Engineers, whose task it was to keep open at all costs the only road between India and Egypt, took three days to get our thirty-five clapped-out vehicles, grossly overloaded and pulling trailers, over the top. They pulled and shoved, unit by unit, with their bulldozers and snowploughs. They must have been very happy to see us going downhill on the other side.

At night we would form a hollow square with the vehicles' headlights facing towards the centre where tents were erected. We didn't cover a lot of ground each day — it took so much time to break camp in the early morning and set it up again before dark. It must have been a cook's nightmare.

Guards were posted and patrolled all night. We were not afraid of a frontal attack but were concerned about being infiltrated by the clefty wallahs [thieves]. If a tribesman gained access to a tent there was danger of a knife in the ribs if the tent occupants woke up during the rummaging of their effects. The thieves were particularly interested in sidearms.

The captain and I had our first bath for two weeks in the River Jordan, not far from Amman. It took some time to remove all the dags that had accumulated in this time — you could almost hear them rattle — toilet paper was not on issue.

Several days later we were back in Cairo and disbanded. It was a very happy unit — for most of us our first, which, it seems, is always the best.

We resented the decision to disband after nine months together in the 'wilderness' where Moses and his mob were lost for forty years. The general opinion was that Moses must have been a lousy navigator!

2

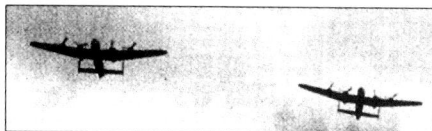

TRAINING FOR INDIVIDUAL MUSTERINGS

Prior to the outbreak of hostilities, Australia had trained up to fifty pilots per year to serve first with the RAF, and then become part of the RAAF. These numbers were inadequate to meet the demands of wartime operations against a Germany which had surreptitiously built up her air force to formidable proportions.

On 20 September 1939, just after the outbreak of war, the Australian government decided to send an expeditionary force of six RAAF squadrons to the United Kingdom. One of these was to be an Australian Sunderland squadron.

At this stage, the RAAF was equipped, almost solely, with Seagulls, Ansons and Hawker Demons — hardly front-line aircraft. In addition, on 6 October 1939 the British Government sought, and was given, the Sunderland aircraft intended for RAAF use. For these obvious reasons, therefore, the plans for an expeditionary force were abandoned one month after they had been approved.

There was, however, the very urgent need to increase the capacity of the UK to wage air war against Germany, both on the offensive and the defensive. A continuing supply of trained airmen was essential for this. On 26 October 1939, the UK Dominions Office put forward a proposal to develop a training scheme in the Dominions, where weather conditions and absence of enemy intervention would facilitate training.

From the subsequent discussions, the Riversdale Agreement was agreed upon and came into force on 17 December 1939. The Empire Air Training Scheme (EATS) had been initiated.

RAAF Hawker Demons — obsolescent at the outbreak of war.

Avro Anson bombers — obsolescent at the outbreak of war.

In fact, the scheme had been first suggested in 1936, but with no result. In December 1939, the parties — Great Britain, New Zealand, Australia and Canada — finally reached agreement.

The Australian contribution was to be about 10 000 aircrew every year — made up as follows:

To be trained in Australia
3 100 Pilots
2 000 Observers
3 300 Gunners (to include wireless air gunners)

To be trained in Canada
80 Pilots (ex elementary flying schools)
42 Observers (ex initial training schools)
72 Air Gunners (ex initial training schools, including wireless air gunners).

Training was to commence in April 1940.

A great effort had to be made to form, equip and staff the many training units required to support this ambitious scheme.

Training, as covered in this chapter, was concentrated on teaching the raw intake the skills of their aircrew category. Each category had different requirements.

The main categories trained were pilots, observers (later subdivided into navigators and bomb aimers) and wireless air gunners (later subdivided into signallers and air gunners). The length of their courses varied.

First, however, it was necessary to make some assessment of the trainees in order to assign them to their categories. This was done at initial training schools (ITS). The length of this course was originally four weeks, but was later extended to six weeks.

The fate of the individuals was decided by the Category Selection Board on the basis of their performance at the ITS. How this assessment was made was somewhat of a mystery to most of the trainees and the training appeared to be about everything not necessarily related to what they might be doing. There were lectures on aerodynamics, some basic navigation and meteorology and plenty of physical training and parade-ground drill. For some reason, considerable attention was given to 'square bashing' under the care of the hated drill instructors. The attention to non-flying concerns seemed, to some, a little overdone. Ross Pearson tells of one ITS experience.

I had come from the Army to RAAF, ITS, at Somers, in Victoria. The inefficiency of my Army unit was in stark contrast to the busy hum of this new service. Here was an organisation which seemed to have a sense of purpose.

On the other hand, some of the course members were keener than others. I recall one 'eager beaver' who was appointed Course Leader. He became a sort of drill instructor, in an honorary capacity. He became quite overbearing in his petty demands for discipline. He even reported several hut members, who subsequently lost weekend privileges. But retribution was at hand.

A small party of 'barbers' attacked his moustache. They were not so vindictive as to shave it all off — half sufficed. Then, to ensure uniformity of appearance, they shaved off one of his eyebrows.

Suddenly, discipline became less severe and our under-corporal became a quite normal human being.

After leaving Initial Training School, trainees followed different paths.

PILOTS

Elementary Flying Training School

The first step for a trainee pilot was to master the basic essentials of flying. This was done at an elementary flying training school, (EFTS).

There was a great gap to close in the experience of the trainees. Few, if any, had flown an aircraft before; indeed, few had even been in an aircraft before. The process of weeding out those with insufficient potential was mainly done here.

Fred Wright recalls an experience at EFTS

I went through EFTS at Temora; we flew Tiger Moths. Nothing very notable happened except that one of our mates dropped his plane in a paddock well short of the landing field. When the medicos arrived, he kept saying, 'Gee, my tinea's bad.' The shock of the landing was too much for him. He fell by the wayside.

Jim McSharry nearly lost his life while training at EFTS at Narrandera.

We had to get up early in the morning to do our flying at Narrandera. It was an inland aerodrome, and in high summer of 1941 the heat made conditions rather unstable from about midday onwards. It was better for inexperienced trainees to fly while the air was reasonably calm. We used to fly from one of the two satellite aerodromes attached to Narrandera.

I had gone solo on the Tiger Moth, but didn't have many hours up, and still had lessons from my instructor — a Flying Officer Wingrove. We had just finished a session of dual instruction and landed, when he said to me, 'Righto, McSharry, I'm taking this plane back. You take that one over there to the main drome.'

I got in the other machine, started it up, took off, and landed at the main drome. Wingrove had arrived before me. He came running out to meet me and said, 'Switch it off! Switch it off! Quickly!' I did so — just where I was. He said, 'Did you look at the oil pressure?' I said, 'I didn't notice anything wrong with it, sir.'

Using the great Australian adjective several times, he told me how lucky I was. They had given me the wrong aircraft to fly back. They had just drained the oil from it and had not replaced it. I had got back on the residual oil in the sump and hanging round the cylinders.

Of course, the engine was extremely hot, but I believe it didn't damage the engine — which just goes to show what marvellous aircraft the old Tiger Moths were.

Service Flying Training School

The courses were originally intended to be sixteen weeks but were reduced, first to ten weeks; then, in October 1942, brought up to twelve; then to sixteen and, finally, to twenty weeks duration.

At the conclusion of the course, those who succeeded were awarded their pilot's wings and were ready to be posted to advanced flying units or operational training units, either at home or abroad.

Tiger Moth

(Dennis Hornsey Collection)

Service Training Flying School, Calgary, Canada. Arthur Doubleday *(third from left, back row)*, Jack Stronach *(Second Row, second from the left)*

Fred Wright tells of some of his experiences at SFTS at Deniliquin, in NSW:
We were training on Wirraways and I was sent up to do some general flying. We used to tackle nimbus clouds, and I dived into one. Then I couldn't hear the engine. I looked at the instruments and found I had no fuel pressure. I had changed over the fuel tank just before this, so I looked again — yes, the cock was turned to On, but I still had no fuel pressure.

So I had a dead motor. I turned, saw the satellite drome, and headed towards that. When I was halfway there I realised I was not going to make it. I looked around and saw a nice little farmhouse. I set the plane down in the paddock in a wheels-down landing. No problems; just a normal landing.

I stayed the night, slept in a nice soft bed, and they put on a wonderful meal for me.

When I got back the next day, my instructor said to me, 'You're lucky the postings have already been made. If this had happened before, you would have got an instructor's posting.'

The first night solo cross-country I did was from Deniliquin to Tocumwal, back to Jerilderie, then Deniliquin.

I was flying along nicely, at 5000 feet, when I got to Tocumwal and looked down to adjust my compass for the next course.

When I put my head out of the cockpit to have a look, I could see things like stars going round and round

in a circle — and I was heading straight for them. 'Something's not right!' I suddenly realised that I was in a spiral dive on to the town of Tocumwal. I had accidentally pushed the stick over and gone into a dive. I pulled out at two thousand feet over Tocumwal.

Flying conditions for trainees in Canada were a stark contrast to those in Australia. The following story by Gerry Judd gives some idea of the differences.
It was snowing, and the aircraft were covered in snow. We still had to fly, although the taxi trails and the runway were quite icy.

We were doing circuits and bumps in Ansons. Trainee

Gerry Judd at Service Training Flying School, Macleod, Canada. Gordon Grellman to his left and Ross Edmonson on his right.

Strips at Service Flying Training School, Macleod, Canada.

Grellman had landed. He had his running lights on, and I landed after him. He was waiting for the signal to take off when I taxied up behind him. I put the brakes on, but the aircraft didn't stop. It just skidded on, and the port engine chewed off his tail.

Grellman was unaware of this, because the wind was blowing quite fiercely and rocking the aircraft, and the noise of the engines drowned out everything. So I had to hop out of my aircraft and race up to his, open the side door, tap him on the shoulder and say, 'Grellman! For Christ's sake don't take off! You've got no tail.'

He said it was the greatest shock he'd ever had. 'There I was, concentrating on getting ready to take off, when I suddenly felt these hands on my shoulders.' He said, 'I didn't know whether it was an angel, or what it was.'

There was an inquiry, and I got three days gaol. But all the blokes had a great laugh about it. At the time,

I was taking out the Chief Flying Instructor's daughter. The normal penalty for a smash like this was seven days, but, although they took the bad conditions into consideration, all the fellows reckoned I got three days because I was taking this girl out.

Ted Eagleton had a close shave at SFTS

I flew a Yale and a Harvard in training in Canada. The Yale had a fixed undercarriage. It was a French aircraft, and the dials in the cockpit were all French. It had a wobble pump in the cockpit for pumping fuel.

I was doing blind flying one day, under the hood, with an instructor. I took off and, the next thing, the engine cut out at about 200 feet. We had to land immediately ahead of us, in a field. It was hairy, but the instructor took over and landed it safely.

There were a lot of accidents in training. I remember, in particular, one trainee who took off in a Harvard at night, with his flaps down. He crashed into a hill and was killed. The flaps added drag and put the nose of the aircraft down.

Eric Cooper tells of an unusual occurrence which occurred when he was Chief Flying Instructor at Wagga Wagga in NSW.

Two Ansons collided and locked together during a cross-country. They locked together in piggyback fashion and the pilot of the upper aircraft was able to land them in a paddock south of Wagga.

When the two aircraft collided and locked together, the pilot and crew of the bottom aircraft baled out. They all parachuted to the ground safely.

One engine of the top aircraft was still operating but neither of those of the bottom aircraft was. With the

Avro Anson.

(Dennis Hornsey Collection)

two aircraft still locked together, the pilot found he had some control through the ailerons and elevators of his own aircraft. With the one remaining engine, he was able to bring the two aircraft down into a very large paddock which had a smooth surface. He landed the aircraft successfully, without further damage.

He was a trainee. I think he was reduced in rank for a while, but he was eventually posted overseas and did very well. He was later killed while riding a bicycle at the aircraft station at Sale.

Both aircraft had their wheels retracted when they collided, and were landed that way. The airscrews were bent, and there was some damage to the engine nacelles, but both were repairable and were put back into service.

William Weller, generally known as Sam, trained in the UK. His comments follow:
Training was too intense at the beginning. Then the routine was changed to half a day flying and half a day of classes. Flying started at about 6 a.m. and the lectures were in the afternoon.

If you didn't go solo in the prescribed period, you disappeared — they put a bowler hat on your head in the course photograph. About 15 per cent used to fail.

When you finished, you were given a piece of paper asking what you would like to fly. The usual choice was fighters. The last thing you would put down was Bomber Command.

I was posted to Little Rissington. We went there by truck. Only two of us could see out through the back of the truck and the conversation was

'Can you see an airfield?'
'Yes.'
'What sort of fighters do you see?'
'I don't know. They've got two engines.'
'What do you think they are?'
'Oxfords or — Ansons.'
'Ansons! They're not fighters!'
So that's what we trained on — Ansons.

WIRELESS AIR GUNNERS
Wireless Air Gunnery School
Trainee wireless air gunners went first to wireless school. Here they were given the basics of radio theory, and learnt to master the sending and reception of morse code. The length of the course was six months.

If they failed at morse, they went off immediately to train as a gunner.

A weakness of training in Australia for service in the UK and Middle East was that the trainees learned little of the British signals organisation, codes and wavebands. Further, Australia didn't use the Marconi equipment which was used in the RAF. Ross Pearson tells of some of the trials of the wireless aspects of training:
I trained at No. 2 W.A.G.S. (Wireless Air Gunnery School) at Parkes in NSW This is well away from the

Pick-a-Back Ansons

Avro Anson at Service Flying Training School, Macleod, Canada.

coast and summer temperatures were very hot and soporific. I can remember being continually awakened by the instructor, who despaired of this AC2 (Aircraftsman Grade 2), who dropped off every five minutes during his radio theory lecture.

I had a great deal of difficulty reading the Aldis lamp. Without any detailed instruction, we were taken to a large paddock where, in small groups, we lined each side. The aim was to send messages, by light signals, from one group to another. The only instruction received came from an, apparently, old man, known to all as Aladdin. His basic, and practically only, instruction was to 'Focus the lamp! You fool!' 'How?' was our natural reaction. But no response came. He might have been deaf, for all we knew.

By trial and error, we fools eventually passed by ensuring that anyone who could manage to read the message called it out to all his group who gathered closely round him.

The monotony of wireless theory and morse was relieved by horseplay in off-duty moments. The natural enthusiasm of young men was exhibited, with all its foolishness and bravado. Ross recalls some of the lighter moments at Parkes:
We slept on beds which consisted of wire mesh tied to a metal pipe frame, with a U-shaped piece of pipe at each end to act as legs. The legs could be folded under the frame, and were attached to the frame by two bolts. The nuts could be easily undone.

It was a popular pastime to undo the nuts and replace the bolts with matches. These would hold the bed together for two to three minutes after occupation,

before giving out. Then the bed would collapse with consequent hilarity and profanity.

This game came to an abrupt end when one of our number, to show his athletic prowess, cleared five beds in a jump from the door, to land on a 'doctored' bed. The bed took off, together with its occupant, and collected five others, who were bruised and shaken. They ensured that the flying trainee was also bruised and shaken.

The same high jinks were performed with the station mascot — a small kangaroo. Some degree of persuasion was necessary to force the roo to bed down in the victim's — who was then at the pub — blankets. Once down, the roo slept until disturbed by the returning airman, seeking to get into bed quietly, after lights out.

One of our community, Bill, had a penchant for sleeping outdoors. He would, on occasions, take his bed outside, to a position about twenty yards from the hut, in the direction of the town of Parkes. He was a heavy sleeper, and followed this practice until. . .

One evening, a small group from his hut, en route back to camp in the early hours, carried the bed and contents about 200 yards down the side of the camp to the Parkes road. A mild joke, and not too far to walk. However, succeeding jokers carefully carried the bed further down the track. A long walk back for Bill when he awoke to find the camp nowhere in sight.

We were subjected to an inordinate amount of petty discipline, and detested the drill of the parade ground. One of our number, a character nicknamed 'Harpo', appeared on the CO's parade with two or three inches of pyjamas showing below his uniform trousers. He became a legend, but came unscathed through his tour with Bomber Command.

In the last month of the course, we were flown round the local district in two-seat aircraft (Wacketts) to practise the transmission and receipt of wireless messages.

To transmit, we had to wind out, by means of a small wheel with an even smaller handle, a trailing wire aerial. This trailed behind the aircraft and confined the pilot to fairly straight and level flying.

I found it unwise to wind in this aerial too soon after our exercise was complete. The pilots, bored with the humdrum flying round of wireless trainees, were wont to take the opportunity to turn the planes inside out. I flew with one, who chased a farmer under his truck and bounced the plane over the top of it.

John Appleton trained in the UK as a wireless operator/mechanic for Coastal Command. If we, in Australia, thought our Fairey Battles were ancient, we were wrong. John's training aircraft were ancient, ours merely very old.

One of the features of Cranwell was a pair of Vickers Valentias — locally known as 'Pigs'. They were twenty years old and had probably done good service in the Middle East, where the RAF used them for troop carrying.

The Cranwell pair had a radio installation at the front of the cabin, with a dozen wicker seats behind. The load of trainees took turns at the radio during a simple exercise with a ground station.

The Valentias, with their open cockpits and biplane wing structure, must have had an operating speed of about 50 to 60 knots. Flying into the sort of winds that the Lincolnshire landscape allowed, often caused them to be almost stationary over the ground.

BOMBING AND AIR GUNNERY SCHOOL (FOR WAG's AND STRAIGHT AG's.)

Australia

Wireless air gunners proceeded to these schools on successful completion of WAGS, and air gunners went to them direct from ITS

The gunnery course lasted four weeks. During this period, the trainees in Australia flew in clapped-out Fairey Battles and used a Vickers gas operated gun. This did not provide suitable training for the turrets and Browning machine-guns of Bomber Command.

Wackett Trainer.

Ross Pearson relates some incidents at the Bombing and Gunnery School at Port Pirie in South Australia:
Gunnery practice was to fire at a drogue towed behind another aircraft. The attacker had to come from behind, dive, straighten out, and let his gunner (hopefully) fire a few bursts into the drogue.

One potential (and he never went any further) gunner aimed with precision, fired with precision, and hit the tail of his own aircraft. His only luck on the day was the tail held together until they landed.

Ming the Merciless was the CO at the Gunnery School at Port Pirie. If he had another name, no one knew it. He could endorse our reports and grant us our coveted single wing brevet or reject us. Our fate was in his hands. He must regard us favourably, or else.

Ming had a sense of fair play for all ranks, and particularly, it seems, for padres.

The usual procedure for church parades in the Services was for the members of each religious denomination to form into separate groups on the parade ground. Each group was then ministered to by an appropriate chaplain, if available. Those without a padre spent the time at rest or on some fatigue duty.

The CO had two padres available — one Roman Catholic and one Church of England. He felt, in justice, that each should have an equal congregation.

So, the instruction was — 'Those to the right will go to the Church of England service. Those to the left will go to the Roman Catholic service.'

No options were allowed, and with an eye on getting commissions at the end of course, and the coveted brevet, no one demurred.

Dudley Hannaford strikes a more sombre note:
I trained on Fairey Battles at gunnery school. We went up alphabetically. It was very cloudy one day, and two planes collided in cloud. They fell like sacks of potatoes and the occupants were killed instantly. I was next up, with two others. We didn't look forward to it.

Canada

As mentioned earlier, flying in Canada during the winter months had its own set of difficulties not met in Australia. This is well illustrated by Jim Dransfield's experience while flying as a pilot at a Canadian gunnery school:

I went to Canada at 18, after I'd done my EFTS at Narrandera. I was commissioned when I graduated and elected to fly as a staff pilot for twelve months before going on to Europe.

One of my jobs was to fly a Lysander and tow a drogue for trainee gunners in Bolingbrokes.

The difficulty was that the Lysander was, in my opinion, almost inoperable in winter. It had a copper carburettor float which was prone to seize up in the extreme cold and cause the engine to cut out. There were many crashes.

The Lysander had a high, strong undercarriage and big spats over the wheels. Consequently, when landed in deep snow, the wheels came to an abrupt halt while the rest of the aircraft kept going — with disastrous results for both pilot and aircraft as it flipped over on its back.

Our flight commander instructed us that, if we lost power, we should get the nose right up, stall the aircraft about twenty feet above the ground and let it drop into the snow with the minimum possible forward speed.

I was caught out one day and had to put his theory into practice. Luckily, it worked for me and I was able to fly the aircraft out later."

OBSERVERS (NAVIGATORS)

Canada

Alby Silverstone gives a good description of both phases of an Observer's course. Quite often, the schools were attended in reverse order to Alby's:

We arrived at Lachine on 11 August, and on the 16th found ourselves at No. 6 Bombing and Gunnery School, Mountain View, which was situated close to the large town of Bellevue, approximately halfway between Montreal and Toronto, on the northern side of Lake Ontario.

Here we were taught to aim and drop bombs, the handling of military guns, including handguns and aerial machine-guns, and the theory associated with each. This included the stripping, cleaning, clearance and loading of each, with particular emphasis on the Lewis and Browning machine-guns.

Classroom work included elementary navigation, aircraft recognition, bomb construction, bomb-loading and the theory of flight.

Initially, the handling of guns and flares was only on the shooting range, but we later went through many forms of aerial gunnery. For our aerial work we used two types of aircraft — the Anson for bombing, and the Bolingbroke, a Canadian version of the Blenheim, for gunnery. In all, I completed twenty-five bombing runs and fifteen gunnery trips in fifty-four hours flying time.

Fairy Battle   (Dennis Hornsey Collection)

Bolingbroke — the Canadian version of the Blenheim.

We moved to No. 8 Air Observers School, Ancienne Lorette, a small village about eight miles [13 kilometres] outside Quebec. Here we learned that the category of Air Observer was now in the course of being changed to Navigator 'B' (the latter for Bomb Aimer) and a separate category of Bomb Aimer had been created. The latter did only a short course in navigation, mainly restricted to map-reading and some lesser parts of the navigation course.

These changes were made in response to the introduction of four-engined aircraft, which required a flight engineer and a specialist bomb aimer. It now seemed we were among the last courses to graduate as air observers.

Navigation school lasted almost six months, from October to the beginning of April. We flew right through the winter, with some four months of snow and ice. The big difference at this station was that it was run by the Canadian Pacific Company. All but the instructors and administrative staff were civilians, working for the CPC. Even the pilots, who, in the main, were seconded from the Canadian Air Force, were considered as civilians, and known as Mr So and So.

The course was very extensive, with classroom subjects consisting of navigation, maps and charts, magnetism and compasses, instruments, photography, reconnaissance, aircraft recognition, meteorology, signals and armament.

The navigation course, both theoretical and practical, was most intensive. It covered all methods of navigation known at the time. It included celestial, radio, dead reckoning, wind drifts, visual map reading and visual sightings. Celestial navigation included the use and recognition of sun, moon and stars, and the use of the bubble sextant. It included the use of great circle navigation, which recognises that the earth is not flat, as portrayed by maps and charts. This is an essential aspect of navigation on long journeys — such as those flying boats and transport aircraft were required to make.

The air work covered bombing and photography, as well as day and night navigation. It was carried out in Ansons.

At first, the snow was a novelty; but then it became a nuisance; and something that had to be tolerated. It was measured in feet rather than in inches and covered everything like a giant blanket. The roads and runways had to be cleared by snowploughs that moved the snow into great heaps on the sides. This gave the roads the appearance of running between huge walls — the runways were rather similar.

It was not possible to heap the snow on the sidewalks in the cities and towns. In those areas it had to be removed by fleets of trucks. The runways were kept even by the use of steam machines, which melted the snow and were followed by rollers to level it.

Everything possible was heated — buses, cars,

buildings, shops etc., and, where possible, buildings were both insulated and double glazed.

The great blanket of snow made visual navigation very difficult and frequent snowstorms often interfered with flying schedules and routines.

The temperature was always below freezing point — and generally held around -10°C. At times, particularly when the blizzards blew, it fell below -40°C

Weather predictions were, generally, very reliable. The weather patterns usually moved across the continent from west to east, so those in the east could see what was coming. Occasionally, however, they went awry.

There was an occasion when we had gone into Quebec for the day — we had the Saturday off. We would normally expect to be back by the time our leave expired at 2359 hrs. An unexpected storm blew in about 4 p.m. (1600 hrs) and covered the streets with many inches of snow. Further, the storm did not abate as expected, but continued for a further three days.

The roads and railway lines became impassable and we were marooned. Hotel rooms were at a premium, but the hotels and inns came to the party magnificently. Although things were very crowded, they fitted everyone in.

It seemed incredible, but for several months the whole St Lawrence River became frozen for more than the hundred miles between Montreal and Quebec City. At that time, the only bridge crossings were at each of the two cities. But during the winter the ice was so thick that motor traffic could drive across the river at many places.

My twenty-first birthday was a very low-key celebration. It consisted of a visit to the city with four other airmen for a steak dinner and a few beers.

As a break from our otherwise crammed program, we sometimes had a weekend leave. We also had a five day break over New Year 1943. During these breaks I visited Ottawa, Niagara Falls and Toronto.

Keith Burns tells of his experience during a practical navigation exercise in Canada:
In the early exercises of the navigators' practical training, we flew two trainee navigators on each flight — one for the outward, and one for the homeward, legs.

I had completed the outward leg, finding the target, either by good luck or by the help of the staff pilot. It was well to the east of our home base — the Air Navigation School at Edmonton. I settled down by the window to take in the view of the prairies and to do some map reading and some calculations.

We passed a town which I thought I recognised — fairly well on track. I calculated ETA base as about twenty minutes. We appeared to be parallel to the Canadian National Railway line, which ran east-west. Suddenly, in response to an instruction from the navigator, the pilot altered course about 45 deg. starboard.

Some minutes passed before the following conversation ensued.

Pilot to Nav: 'Do you know where you are?'

Nav to Pilot: 'We will be over Edmonton in six minutes.'

Pilot to Nav: 'Bullshit! We are off the top of the map.'

Nav to 2nd Nav: 'What do I do now?'

2nd Nav to Nav 'Give him a reciprocal course and give me a look at your log.'

A quick look at his log showed that, in converting course true to course magnetic, the nav had applied the correction the wrong way. In that part of Canada the variation was 26 deg. west. Any wonder we were off the map?

Noel Gilmour also had a problem:
I got lost on one of my first training flights. Well, not really lost, because these flights just kept going around the eastern side of Edmonton and one soon got used to the ground features. For example, there was a conspicuously shaped lake that could be seen for miles — it was a handout for a navigator.

The pilots were civilians and they just flew navigators round and round and round. They knew the country backwards.

There was no Gee, so I was basically flying around, 'navigating' by known ground features.

Noel also had some comments on his training at Bombing and Gunnery School:
A major problem was the hand-held F24 camera which we had to use.

As a record that the 'target' had been reached on an exercise, you had to take a picture from nearby, have the pilot do a steep turn over the place while you took another picture, and then fly a little way off to take another picture, before setting course for the next target. We had five navigators on board and five targets to reach, with two steep turns over each target.

During the steep turns, the camera was too heavy to hold, because of the 'G' force of the turn. So it had to be put on the floor while you turned the handle for the next picture. There was no alternative to putting your head down and grinding away.

By the time the five targets had been found and photographed, all the navigators were as sick as dogs.

Australia

Doug Nicol trained as a navigator and, while training at Cootamundra, had an unusual forced landing:

We were on a navigational exercise, flying from Cootamundra, in NSW, to Nhill, in Victoria, and return. That was about 700 miles or so, as the Anson flies, and just about its limit. We had six trainee navigators and one sergeant pilot on board.

We'd got to about Junee, on the way back, with about thirty miles to go, when one motor cut out from lack of fuel. The other was pretty low as well, so the 'driver' decided to put down in a hurry. I had the task of winding down the undercarriage — it took thirty-six turns of the handle to do this, but I didn't need any urging and got it down in about ten seconds flat.

Our sergeant put us down in a paddock, missing a ditch which ran across it and a wire fence on the approach. I swear, to this day, it was like Moses and the Red Sea the way that fence opened up and let us through.

We landed right in the middle of a flock of sheep. There was an old shepherd there — he even had a crook, which he waved around wildly — who carried on a treat. He was far more worried about his wretched sheep than thankful for our lucky escape.

As it was, he ended up with more on his plate than he anticipated. One of the boys was sent off to phone base. The next thing was the arrival of another Anson — with the CO and more petrol on board. We filled up and both flew off without more ado — leaving the shepherd, his crook and his sheep to their own devices.

PASSING OUT PARADES

The presentation of wings was the occasion for a ceremonial parade, at which the trainees were the guests of honour. Peter Matthews presents a comment on his passing out parade, which was quite memorable for him — as well as for one of the drill sergeants:

The Air Officer Commanding Eastern Area of Canada took the salute at our passing out parade on Prince Edward Island, at which we received our wings.

It was quite a good parade — there would have been some thousand airmen and airwomen on the parade.

The air vice-marshal was on a big dais an the side of the parade ground. We had to turn in towards him to receive some words of encouragement and congratulation.

At the end of the ceremony, the drill sergeant confused the whole operation by turning us the wrong way. Instead of turning us left to march us away, he turned us right and we were facing the people who were supposed to be following us, but who were now in front.

The poor old flight sergeant was covered in embarrassment. I think his career came to a sudden halt.

3

ADVANCED AND OPERATIONAL TRAINING

Most of the air crew personnel who arrived in the UK, while trained for their various musterings, lacked any experience of working as part of an integrated aircrew. There were some exceptions, because there were some Operational Training Units in Canada and, later, in South Africa. But the majority had to pass through four phases before commencing 'serious' operations. These were: personnel reception centre (PRC), advanced flying unit (AFU), operational training unit (OTU), heavy conversion unit (HCU).

PERSONNEL RECEPTION CENTRE

On arrival in the UK, Australian aircrew were immediately posted to the personnel reception centre. This was a holding centre for aircrew awaiting posting to advanced training. At times, the wait could be several weeks and aircrew became bored by the delay, particularly as no reasons were given.

Gerry Judd was one who avoided the boredom:
The personnel reception centre was in Bournemouth at the time, and the town used to have an afternoon tea dance. It was very popular, the ladies used to get there quite early and sit close to the dance floor. You had to get a table near the dance floor, or you were sunk.

My friend, Harry, once again, found out a lurk. We were doing aircraft recognition lectures at the time, and they didn't finish till 4.30 p.m. The dance started at 4 p.m. Harry found that the louvres could be pulled out in the toilet near the lecture room. I was quite slim in those days, but Harry was stocky. He said, 'You get up through there.' Despite my doubts, he persuaded me.

He said, 'Throw your books through, and I'll bring everything else. You go down and get a table.' So, at about four o'clock, I'd excuse myself and go to the toilet. I'd go in, climb up, and Harry would push me through. I'd collect my gear and get down to the Pavilion. There, I'd get a table near the floor, and tip a few chairs around to reserve it. When the other chaps came down, we were firmly ensconced. This used to upset them no end. We were right on the spot and could pick the prettiest girls.

ADVANCED FLYING UNITS

Flying conditions in Europe and the UK were vastly different from those in the countries in which the new airmen had been trained. They needed the opportunity to adapt to the different weather patterns, topography, etc.

This was provided, together with training in more advanced skills, in advanced flying units.

Some categories of Australian trained aircrew had the difficulty that they had been trained on equipment which differed from that used in Britain. This was particularly true of wireless air gunners and gunners. There had been little, or no, training on modern turrets, and aircraft recognition had concentrated on Japanese types.

Navigators suffered from lack of training on the Dalton computer, and had no practical experience of astrocompasses, drift recorders or distant reading compasses. When you got your wings you knew the basics of your trade — rather like getting your bachelor's degree — and were expected to master the more advanced bits later, whether in Europe or Australia. The specifics depended on the theatre and the equipment used there.

Bruce Otton trained on Oxfords at AFU:
Oxfords were generally regarded as a somewhat difficult aircraft to fly, mainly because it was very difficult to do a three-point landing with them. The best way to land them was to do a tail-down wheeler.

The pilots who had trained on Wirraways in Australia seemed to qualify on Oxfords more quickly than those who had trained on Ansons. In fact, it took me about six hours to do the conversion before I was sent off to go solo. Funnily enough, I enjoyed flying Oxfords better than any other aircraft, because I really felt master of it.

My instructor had been a Tempest pilot. He was a wild young man who said he used to do patrols up and down the North Sea. He and I used to have competitions to see who could do the steepest turns.

When we got into night flying, he used to deliberately make life difficult. I remember we were coming back to base one dark and stormy night, when he said, 'You're on your own.' Much to his chagrin, I managed quite well.

Blind-flying was a hazardous exercise for a number of good reasons. Loss of the airspeed indicator was a distinct worry at any time — not so much in normal flight, perhaps, but the margin above stalling speed when landing was very small, by day or by night:

Bruce Otton:
The most important thing you feel, when you're learning blind flying, is that the airspeed indicator is absolutely vital. No instruction was given on techniques to use if it packed up.

We were doing a cross-country one night, when, suddenly, I saw that, according to the airspeed indicator, I had no airspeed.

'What'll I do? We've got no airspeed. Good Grief! We're going to fall out of the sky' — until more rational thinking got under way. I thought, 'Now wait a moment. If the engines are set at the right boost and we have the right propeller setting, and the rate of climb indicator isn't showing us going down, and the altimeter is steady, we must be flying and have the right airspeed.'

So I pressed on. When I got back to base, I decided to do a flapless landing with engine on — and there we were going along the runway, and touching down.

Bruce Otton goes on to tell of a training crash at AFU:
While training on Oxfords at Kidlington, I was called up to the chief flying instructor's office about two hours after I had landed from an exercise.

He asked whether I had noticed any strange ' movements' in the aircraft when I was flying. I said 'No.'

Another trainee and an instructor had taken off in the aircraft I had used, about five minutes after I left it. They had crashed and the trainee had been killed.

The instructor had survived the crash, and, when rescuers arrived, was leaning against a fence and about to light a cigarette. When someone came over and said, 'Are you all right?', he just dropped dead. There were no marks on his body so it was attributed to shock.

Arthur Doubleday tells how he was nearly killed at AFU:
The first time I flew in England gave me my worst experience. It was supposed to be a familiarisation flight. It was 1941 and crews were needed badly.

The chief flying instructor said to the two of us, myself and a New Zealander, 'It's not very good, I know. But just go and do one circuit. If it's no good, come in and land.'

So, we both took off. That was the last I saw of the drome — we were in cloud at 300 feet.

Then we got a message that, if we considered conditions impossible for landing, we should steer a set course and bale out. The instructor, a sergeant pilot, said, 'What do you think?' I was a pilot officer. I opened the window and then said, 'Give it one more go.' He agreed and said, 'I'll fly it and you call the altitude.'

We got down to 150 feet and, there below, was an airfield. We landed successfully.

After we'd got down, we found that we'd flown over a 500 foot ridge while in the cloud, and should have killed ourselves. As it was, the New Zealander had crashed and blown up.

Fred Wright also had to fly in bad weather at AFU:
The weather was bad and the instructor told me there was a lot of cloud coming across from the east, but I should go up and keep an eye on the little open pocket over the airfield.

So I took off and was enjoying myself, until I looked down and found 10/10ths cloud. Knowing that the cloud was coming from the east, I turned west, and came out of the cloud. But there were funny white things everywhere.

I knew our field was on high ground and felt I would hit a mountain if I looked for it while flying blind, so I put the plane down in a paddock. Then the mystery of the white things was solved. They were snowflakes. I'd never seen snow before.

Loss of visibility was only one of the difficulties that bad weather presented. Noel Gilmour, a navigator, explains:
At Halfpenny Green, we did two cross-countries, one after the other and, on the last day, there was absolutely foul, unstable weather all over England. We were to do two trips, of two and a half hours each, that night.

We did the first trip in about three hours. It was a dreadful trip, the plane bucked and tipped all over the place. It was difficult to do anything.

Arthur Doubleday's Commission

It was nearly midnight when I got back and we all went to the mess for dinner, although I was still not feeling very well. I walked down to the Nissen hut and was violently ill.

The met people came to the second briefing and said, to our joy, 'This will be much the same.' We said, 'Better scrub it.' but they didn't.

All I could do on this trip was give the pilot courses based on the winds I'd found on the first trip. For the rest of the time I had to hang on to my table. I was too sick to do anything else. I couldn't have cared less if the aircraft had crashed.

OPERATIONAL TRAINING UNITS

Air Chief Marshal Sir Edgar Ludlow Hewitt, who headed Bomber Command at the outbreak of war, realised that operational crews were in need of better training for the hazards of operational flying. He initiated steps for the formation of operational training units.

These became an important prerequisite for operational survival and efficiency. It was here that crews were chosen, and individuals welded into a combat team — although Flight Engineers did not join crews at this stage. They joined at the Heavy Conversion Units.

The OTU in Bomber Command was, essentially, where

crews were formed and given the opportunity to integrate into a team, while adapting to higher performance equipment. The training exercises were meant to be closer to the sort of thing that would be met on operations, but were by no means operations in any real sense. You, hopefully, came out of it in a crew which knew one another's strengths and weaknesses and with the skills from their basic training expanded to meet a more demanding situation. But operationally experienced, you were not.

A big problem was aircraft availability. Indeed, many accidents in training were due to aircraft being 'clapped out'.

Including Dominion personnel, 5237 RAF officers and men were killed, and a further 3113 injured in accidents during 1939-45.

An OTU course lasted about fourteen weeks.

Trials And Tribulations Of Cross-Country Flights

After familiarisation with the aircraft, by circuits and bumps, simulated emergencies and non-standard flying training, crews were ready for cross-country flying.

'Ready', was probably not the right word. They were sent on these trips in various stages of readiness and experience. The situation is explained in more detail by Alby Silverstone:
The aircraft's hydraulic system failed during the flight

and we lost the ability to raise or lower the undercarriage and the flaps. These are rather vital to the landing operation.

No one thought of the emergency pump — although we should have, particularly Aub [the pilot], who had almost knocked himself out with the handle.

It was possible to wind the undercarriage down, but once down it could not be wound back up. We had decided to make a flapless landing, and I wound the wheels down in preparation, and then strapped myself into the second pilot's seat.

Now a flapless landing was always tricky, as it entailed a fast, low-level approach and a high-speed touchdown. They were always difficult to judge and, with an inexperienced pilot in an unfamiliar aircraft, a bit dicey.

We made two attempts, which were much too fast, and it was necessary to accelerate each time, make another circuit, and try again.

Then the control tower asked if we had tried the emergency system!

This was nearly as bad as the flapless attempts. The flaps had to be selected as up or down, as desired, and then pump the handle until the required position had been reached. The handle was where the second pilot usually sat, so I had to fold up the seat, which was collapsible and suspended above the aircraft floor, some feet below.

To operate the pump, I had to stand, unsupported, on the edge of the upper floor and pump the handle up and down until the required position was reached. The routine started as the plane approached the runway.

As before, we came in too fast, then too steep, and each time I had to pump the flaps down and then pump them up again. Each operation involved some hectic pumping.

It involved several approaches before we finally managed to make a landing.

The whole thing was very dicey as, in my precarious position, I had no support, and a bad bump or an overbalance meant a fall on to the dashboard or to the plane floor some feet below.

Bruce Otton had an experience:
They sent us off, one night, each with a companion pilot, a trainee, on a cross-country in the direction of London. We got mixed up with a raid on London and were shot at by our own anti-aircraft fire.

The drill was to flash the letters of the day on the signalling light beneath the aircraft. But when we were caught by a searchlight, close to the balloon barrage, and the anti-aircraft was pumping away, we weren't too sure what to do.

So we started pounding out the letter of the day, diving, climbing and turning until we got out — in one piece.

Scattered locations of OTUs

OTUs were located over the greater part of the old British Empire. George Gray served on one in Kenya:
I operated, and was hospitalised, in the Middle East. I finished my tour just before the second battle of Alamein began and was sent to South Africa to do a course to become senior navigation officer at 70 OTU at Nakuru in Kenya. I was supposed to be having a rest period for a year before returning to 454 Squadron.

We started out to train South Africans under the Empire Air Training Scheme. Then we received large quotas of Englishmen, Australians and New Zealanders. We even had a character from the Argentine and someone from Fiji. It was a vary hard-working OTU. We had a day off every tenth day.

We worked from 6 a.m. till 8 p.m. We started at 6 a.m. and flew till noon. The thunderstorms started about midday, so we didn't fly again till about 4 p.m., when we'd go on till 7 or 8 p.m. We were churning out sixteen crews each fourteen days and sending them up north.

I recall that, at Churchill's insistence, we had a group of six or eight crews of Turks. They did a full RAF course. Usually, the commissioned officers were navigators and captains of aircraft. The other crew members were warrant officers. They were all from the permanent Turkish Air Force.

As Navigation Officer, I devised the courses. They consisted of a variety of exercises to make crews aware of, for example, care about height. This was necessary because we were operating at 6000 feet above sea level, and had Mount Kenya to the north of us and Mount Kiliminjaro (about 20 000 feet) to the south of us. There was also a high plateau, which went up to 12 000 feet, to the south. It was covered with timber. Lake Victoria was to the west. We used to send crews over there on navigation exercises.

We sent one of the Turkish crews on an exercise to the south, warning them of the plateau at 12 000 feet. They were to go round Mount Kilimanjaro, pick up certain points, drop flares, etc., and come back to base. It was about a four-hour trip in a Baltimore.

However, we lost them. They went off the plot. I had to organise a search pattern to cover the 600 hours allocated to a search, if necessary. But we didn't find them in the 600 hours. This was quite understandable, as the area was subtropical and had subtropical

Blenheim bomber.   (Dennis Hornsey Collection)

vegetation. Nakuru was one mile south of the Equator.

We closed down the search. Then signals came through from London. Restart the search. We can't have this catastrophe for Turkey.

A native operating at the snowline during his hunting found the wrecked aircraft on the snowline of Mount Kenya. The ground party reached the plane. It was a plus for us.

Because of the language problem, we had to be very careful in briefing the Turks. I recall that the interpreter who was giving the navigation instructions was a bootmaker from Chicago. His English was typical of Chicago double talk. I'm sure that instructions trying to explain magnetic deviation, etc. were interpreted in rather unusual ways. However, that's just a guess.

We carried on at Nakuru for quite some time, still working with Turkish crews and, mainly, RAF and RAAF, until the instruction came for us to move to Shandur, on the Canal. They were building another base there, an OTU. It was nearer where the action really was.

We took formations of aircraft north, along the Nile — from Lake Victoria, up the Nile, through Juba and Khartoum, Luxor to Egypt. One hundred and five aircraft were despatched.

Our CO — Wing Commander Smith DFM — said, 'Gray. Report to the tarmac. Aircraft number so and so.' I said, 'That's a Blenheim Mark I, Sir.' He said, 'Yes. We're going to lead our formation in a Blenheim Mark I.

Go and draw a pilot's parachute.'

I did an hour and a half flying circuits and bumps as a pilot, with him in the number two seat because, he remarked, 'If anything happens to me you'd be able to get down.' I thought this was a good idea but...

The Scottish wireless operator, James McBeath, and I looked after Wing Commander Smith on our journey north. It was quite an experience of how primitive conditions were. There were two airfields where we were refuelled from four-gallon petrol tins. 'Dinkers' (locals) passed a tin up from under the wing and tipped it into a filter. Then they passed another, and another, until 400 gallons were lodged in each tank. They took some time.

At one stop — the Nile was very fast there — the wireless operator and I found we could go across a wooden bridge over the Nile, to a sort of rest house on the western bank.

There we managed to find a warrant officer who could provide us with a bottle of Scotch for the wing commander's edification. We came back across the bridge, which was a bit rickety, but had been used to bring in supplies for the Ethiopian war.

Someone said there were supposed to be elephants on that bridge, and we said, 'Christ! We didn't see any'.

It transpired that five elephants had got onto the bridge. They couldn't hunt them off in case they panicked and broke the bridge. So they just had to let them go. They all just walked off on the western side before us.

We confirmed that there were elephants there as we took off next day. There were the five elephants, just near the end of the runway — a male and a couple of females and young.

North of Wadi Siebena, the aerodrome for Khartoum, round about Luxor, we flew through a cloud of locusts — they covered the sky. They obscured the windscreen and we had to use all sorts of devices to clear it. Smith brought it down at the nearest aerodrome and we had to get the engines (radials) cleaned out.

The pivot head wasn't blocked, but the plane was covered with green slime, which had to be cleaned off with 100-octane petrol. So I believe the Bible story of the plagues of Egypt.

We set up an OTU at Shandur, on the Gibber Lake, between the two canal ends which make the Suez Canal.

I had the job of setting up a simulator for navigators. It was built from eight nose-cone ends of Baltimores that had been damaged and discarded. They were cut off and set up in a large hall with no windows. They were lit up like a Baltimore compartment, and we had photographs on a screen, e.g., crossing the coastline, a photograph of a ship. It was an attempt to devise a, pretty primitive, simulation training set-up in the desert. It seemed to work well by the time I was posted to 454 Squadron as squadron bombing leader.

Ron McCathie tells of a problem with training at OTU at Nakuru, some 200 miles north of Nairobi:

The aircraft were Blenheim light bombers. The tyres were not designed for use at 6000 feet above sea level. The usual landing speed of a Blenheim was seventy miles an hour, but, at this altitude, it had to be increased to 120 miles an hour. Any rough landing resulted in a burst tyre.

We were lucky, and had no casualties, but the rubber department had difficulty maintaining serviceability.

Crashes

Losses due to crashes were quite high at OTU the aircraft were overworked and, too often, were undermaintained. In addition, crews were learning the ropes and were quite inexperienced.

Dudley Hannaford survived a crash at OTU and tells the story.

I was flying in a Wellington. We were coming back from an exercise when one of the engines gave trouble, and then cut out as we were on the landing approach. Then we got a red light (a warning, by signalling lamp, that the runway was obstructed), which meant we were not to land, and we had to make another approach at about 600 feet. As we did this, the other engine started to play up and the pilot lost control of the plane. He said, 'For goodness sake (or words to that effect) brace yourselves, we're going in.' So, we did.

Thankfully, it happened right over the runway, but we hit the ground at 150 mph and skidded 150 yards. the engine was still running when we hit and, of course, knocked the propeller around.

As soon as the plane came to rest the engines went up in flames. There was a breeze blowing, and it blew the flames right across the fuselage. The Wellington was fabric-covered, so it wasn't long before the plane itself was burning.

Wellington Bomber.

A crashed Wellington at an OTU, Lichfield, Staffordshire.

I tried to get out through the astrodome — the navigator had got out just before me and was okay. I was the last to try. As I did, the flames flashed across, and I got the backs of my hands covered in third degree burns and I was burnt on the face.

I thought, 'I can't get out here,' so I went back into the fuselage and tried to get out through the front. But I couldn't get past the second pilot's seat, and couldn't undo it to collapse it.

I could hear the crew calling out to me, and wondered why they didn't get away, because the plane was going to blow up at any minute. I could hear the rounds popping off, and honestly thought this was it.

As I came back up, the flames were following me. I thought, 'What can I do? I'm stuck.' Then, lo and behold, the rear gunner had the presence of mind to climb up on the wing, hop over the top and undo the seat. When he did that, I was able to get out through the front.

All the senior officers were around, outside, and they couldn't catch me. I ran as fast and as far from the plane as I could get. The CO finally caught up with me and said, 'Come here young man. Let's have a look at you.' I said, 'I'm okay. I'm all right.' He said, 'Come on now. I'd like to look.' I didn't feel any pain at the time. But when I got to hospital I felt all the pain in the world.

I was the only one injured, everyone else got out unscathed. As soon as I got out of hospital I was told I should get back on to flying, so I did.

This time, the pilot had the DFC. I must say, he turned to be a rather hairy flier, and I wondered what I was doing in a plane with him. I had to check the wireless gear, as we always did, and he called me up on the intercom and said, 'How are you doing down there? When you've finished, come up and we'll have a look around.'

So I came up, and sat in the second pilot's seat, and we had a stooge around. I can recall, very vividly, there was a group of trees with a narrow break through them. I couldn't believe it, but the pilot went for the gap, just for fun. He would have to tip the wings to get through. As he was approaching, he looked at me, and when he saw my face he said, 'Perhaps you'd like to go back now?' I said, 'Yes. I think so.' Thank heavens, he did.

Arthur Dutch's diary records some events which led to fatalities while he was at 170 OTU at a Silverstone satellite called Turmeston:

Another bad day. About midday we learnt a bomb aimer was killed at Silverstone (baled out without his chute). Rest of crew made it OK. Later, one of our kites on a gunnery trip over the Wash ditched. Six out of seven were saved — the WOP was not seen. Last chap got out about thirty feet under. The rest were picked up by a destroyer after about twenty minutes in the dinghy. The Wimpey went down in four seconds.

Bruce Otton had a close shave at OTU:

I recall an occasion when, on my first or second night flying exercise, I had a 'screen' [an instructor who had completed a tour and was 'screened' from further operations for a while] with me. We hopped into the aircraft, did a quick instrument check, and taxied out. We had to select about thirty degrees of flap for take-off. I selected 'down' and got a bit more than thirty degrees.

To correct it, I lifted the lever up and put it back into the central position. This should have given the correct flap setting.

We then taxied on to the, fortunately, long runway, lined up, and off we went. We bored on and on and on down the runway and, eventually, hauled it off at the end.

The night was pitch black and no horizon as we staggered off and climbed to 800 feet, where we normally raised the flaps.

This I did, but had an immediate sinking feeling. I looked at the altimeter — it was going round in reverse, and the rate of climb indicator showed that all was not well.

The screen and I hauled the stick back with some difficulty, and we got back into a climbing attitude. The mid-upper gunner said, 'Did you see that tree?' and the wireless operator said, 'What about that roof we just missed?'

We assumed that, because the aircraft was so old, the notches for the flap lever were well worn. When I selected the neutral position, the flaps had slowly run down to the full flap position. The resulting drag then made it hard to get off, and raising the flaps made the nose drop.

Ron Goode's experience at OTU included a crash in a Spitfire:

I was eighteen stone when I was called up and I never got weighed after that. I was in England and instructed on Miles Masters for two years before going on to Spits in Shropshire. while at OTU there I had a

Spitfire seat 'die' on or, rather, under me.

We were practising dive-bombing when the seat just collapsed under me.

I tried to pull the stick back, but the seat had moved forward and there wasn't room for the stick to come back. The only alternative seemed to be to 'bunt',[1] get negative G and lift myself and the seat.

Fortunately, I was strong in the legs and was able to brace myself, hold the ratchet on the seat and lift it. I was right on the deck when I managed to pull out. Everyone thought I'd gone in.

I considered baling out — but then I thought of my twenty stone on a twenty-foot canopy parachute; I'd break my leg.

I pulled into the circuit. I couldn't call up for landing clearance — as usual on Mark V Spits at OTUs there was always something u/s; today it was the turn of the R/T.

It was as I turned into the circuit that I realized that the rudder was jammed. When I pulled the seat up the armour plate at the back of it hadn't come with it. It was jamming the control cables.

A Spitfire is always landed with full left rudder, otherwise it just swings round hard to the right—always! With the rudder jammed I couldn't control it

1. Bunt — An 'outside' loop. Instead of pulling and holding the stick back, as in a normal loop, the stick is pushed and held forward. The aircraft dives, passes through the vertical and horizontal with the pilot on the outside of the circle. Due to the increase in speed and the fact that gravity acts in the same direction as the centrifugal force of the loop, the negative G forces become very high. Aircraft are designed to withstand higher positive G than negative G forces so a bunt is not a recommended manoeuvre.

Spitfire of Queen's Flight visiting Imperial War Museum, Duxford, Cambridgeshire.

with the brakes either. So I knew what was going to happen.

I brought her in very gently at 90 mph I wouldn't have broken an egg with that touchdown. As soon as the wheels touched I switched off the ignition and the petrol. We just went round and round on the grass and then went quietly up on the nose in the middle of the field. I got out quick smart; Spits had a bad habit of catching fire in these situations. I got a green endorsement[2] for bringing the aircraft back.

Training To Survive A Crash

Harry Brabin, a wireless air gunner on 102 Squadron, focuses on another aspect of operational training — ditching drill:

Learning to abandon an aircraft, in case we landed in the sea, was a manoeuvre our crew excelled at. We practised sitting in our crash positions, in the fuselage, behind the engineer's position. We sat on the floor and braced our feet against a strut, and our backs against another strut, with our hands at the backs of our heads, and our knees bent, so that we would not be tossed around on impact. It was also a good position for praying.

To leave the plane, we got out of the mid-upper escape hatch, carrying radio transmitter, rations etc., slid down on to the wing and inflated the dinghy, all in eleven seconds.

We seemed to move as one man as we ran down the fuselage, grabbed a strut, swung our legs out of the hatch, and slid down outside the aircraft with only inches separating us.

Crewing Up

Individual categories were moulded into operational crews at OTU. In most cases, airmen had a degree of choice as to which crew they joined.

A range of methods was followed. Ted Priest, a rear gunner, explains:

I was posted to 10 OTU at Abingdon, where we crewed up. It was a funny set-up, because the pilot that I finally crewed up with didn't want an Australian. He felt they wouldn't have had enough training and wouldn't know what they were up to.

Fortunately for me, the Irish bloke the skipper had originally chosen, dropped out. He was then told that he had to have an Australian.

The pilot and navigator had, apparently, decided while in the pub, to crew up. They came to my hut and asked me whether I'd join their crew.

This was my first tour crew, and it was a very happy one.

Max McVicar, a pilot, explains how he crewed up:

There were two Australians in my crew, myself and the wireless air gunner. I had a navigator from Middlesborough, a bomb aimer from Southampton, a mid-upper from Perth in Scotland and a rear gunner from London.

I crewed at the Moreton-in-the-Marsh OTU All categories, except the flight engineer were already there.

Over a period of a week or so, I wandered round and crewed up. I picked the crew by just talking to them. One gunner knew another gunner, and so it went.

My Australian wireless operator was commissioned before me, but that didn't cause any problem. The captain of an aircraft often had commissioned officers in his crew.

Bob Henderson, another pilot, explains how he arranged crew allocation:

I was at No.810 OTU at a place called Whitchurch Heath, in Shropshire, flying Armstrong Whitworth Whitleys.

Before we started the training, we had to be allocated three crew members. They were navigator, wireless operator and rear gunner. The other crew members — mid-upper gunner, flight engineer and bomb aimer — were allocated later, when we went to conversion unit at Lindholme.

At Whitchurch, there were equal numbers of all categories. I, as senior pilot, was asked to ensure that the crewing up was done in an orderly fashion. So I let them all mill around, it didn't look to me that any particular chap was any better than any other.

I let the other pilots pick their crews — what was left were mine. I don't regret what I did.

Jim Tudberry reveals how he, as a pilot, chose his crew:

Everybody was in a big hall, and I just went round and said, 'I'll have you as a navigator, you as a bomb aimer,' etc. They just looked efficient to me. There was a lot of luck in it — a terrible lot of luck.

Bruce Otton adopted a variation in his crew selection:

We went to Lichfield, late in July 1944. We were assembled in a hangar and told that there were members of all aircrew categories there and we had

---

2. Green endorsement — if a pilot did something particularly good, or particularly bad, in handling an aircraft the powers-that-be made a notation in his log book The notation was in green ink if the act was praiseworthy, such as saving an aircraft in an almost unflyable state, as Ron Goode had done. If the action indicated gross incapacity or his inability to get his finger out, it was made in red ink.

twenty-four hours in which to crew up. If we didn't crew up in that time, we would be forcibly crewed up. We were also advised that any aircrew member could approach any other aircrew member to crew up, and any aircrew member could reject an approach by another aircrew member. I recall that we went back to the huts where all the pilots were accommodated, and my mate said, 'What an embarrassing position! Gee. Do you feel like going up to somebody and saying, "Will you fly with me?" What if he turns around and says he doesn't like the look of you? Look at me. I'm only a little short arsed bloke, and young as well.' He did, in fact, look young for his years.

However, as luck would have it, I went down to the mess and ran into Bob Molyneaux, who had been on course with me in No 32 Course at Deniliquin. He was having a drink with Bill Brett, who had been with him in the bank in Victoria. Bill was a wireless operator with our group into Litchfield. So Bob introduced me to Bill and, after we'd had a drink, I said to Bill, 'Are you crewed up yet? Are you prepared to take a risk with me?' 'Oh. I guess so.' So we got together.

As we were walking back towards the huts, we saw a short, moustached chap coming towards us. Bill said, 'Gee. That chap looks familiar.'

We stopped and said, 'Who are you?' He said, 'I'm Ron Mason.' He also had been scrubbed from Benalla, and had gone to Canada to train as a bomb aimer. He agreed to join us. I was concerned, though. I was then only twenty, Bill was twenty-five and Ron was about thirty.

We decided we'd better move fairly smartly and get a couple of gunners, as they had the reputation of being somewhat wild characters. We felt that, if we didn't get moving, the best would be picked up by other pilots. So we walked through all the huts where there were gunners and looked them over, as if they were a herd of cattle, or girls at a dance.

We went back to the first hut. There was Ross, sitting on the base of his spine, reading a book. Geoff Symonds was ironing his tie, which was pretty typical of Geoff, who was pretty smart.

'Howdy, fellows. How'd you like to join us?'

They looked us over, and asked to see my log book. They couldn't see anything that was terribly detrimental, and then asked to see the log books of the others. They then decided to crew up with us.

We needed a navigator. We went through the navigators' huts, but we didn't see anyone who appealed to us, on first appearances. Then Ron Mason said, 'Hey. I think there's a chap, an officer, who was on course with us, from WA. Why don't we test him out? We

Bruce Otton and crew, 466 Squadron RAAF, Driffield, Yorkshire. Front row, from left: Ron Mason (bomb aimer), Bruce Otton, Mal Canavan (navigator). Back row: Bill Brett (wireless air gunner), Ross George (air gunner), Jeff Symonds (air gunner), Ken Johnston (flight engineer).

tracked him down the next morning. He was thirty-one or thirty-two. We checked his log book and found he had topped his course at Mount Gambier. We reckoned he was pretty good and developed an instant rapport. He was ready to join us.

Things looked very good. Unfortunately, the Powers That Be decided to give the navigators a new test. They had six minutes to work on a set of navigational data, from which they had to find the wind and set a new course.

Possibly because he was older than most of his companions, our potential navigator did not complete the exercise in the given time. He was scrubbed off that course and held back to a later one. We were simply told that we had lost him, and were given a navigator who had not been crewed up before.

Jack Robertson's experience in crewing up in the Middle East was quite different to the UK experience:
When I went out to North Africa, I was a second pilot. It turned out that quite a number of crews had time expired after the North African campaign, and they were a bit short on second pilots.

A lot of the crews that came in at my time had ferried Wellingtons, either to the Middle East or to North Africa. Some of them had flown Hudsons, and just walked in and said to the CO, 'I'd like to join your squadron.' There weren't quite the same rigid procedures as in the UK.

I found myself with the former skipper of a Hudson. He was part time-expired. I flew with him for a couple of months and gained some experience. We had two flights operating then, one out of Corsica and one out of Italy. Both flight commanders were Canadians.

The flight commander in Italy had a number of Canadian first pilots. A Canadian broadcasting crew had come out and flew with a number of these pilots, and the flight commander, to show them the merits of the Marauder.

Unfortunately, they crashed on landing, with a number of fatalities, and were short of first pilots.

Very smartly, a number of us were sent down to a place called Telergma, about ninety miles south of Constantine, in Algeria, where the Americans had a Marauder training outfit. We were converted to first pilots in a hurry.

When I came back, I picked up an Australian second pilot who had come up from Cairo — Ivor Duffel, a new navigator — a Gloucestershire man — a very solid bloke, and three partly tour-expired aircrew — a Scot, a Lancashire coalminer, and a Southern Rhodesian.

Bullseyes

Towards the end of training, the crews were required to fly on a bullseye — a cross-country flight simulating operational conditions.

Alby Silverstone had a lucky escape on one of these:
A bullseye was a major training flight under operational conditions, and usually entailed some 4-500 training planes flying, en masse, to a point close to the enemy coast. This usually coincided with a full Bomber Command raid to some other point of enemy territory so that, in addition to the training benefit, it acted as a diversion to the raid.

On our bullseye, we were to fly to a point close to the Dutch coast. But before we got there, the weather over England deteriorated and we received an early recall. The weather worsened rapidly over most of central England and a general diversion was radioed for all aircraft to land on east coast-stations. Somehow, we didn't receive it.

Our trip back was made at the operational height and was well above cloud. We were still above 10 000 feet when we arrived over base and, on calling up for landing instructions, we couldn't receive an answer.

After some minutes we concluded, quite wrongly, that our radio was u/s and decided to descend and contact base visually. It was now getting dark and Jim, our navigator, voiced the opinion that, as we had no idea of ground visibility, we should take a line of descent that would take us well clear of a range of low mountains. To do this, I set a Gee [semi-radar navigation aid] line that followed the proposed line of descent on the screen so that Jim could follow it, and I moved into the second pilot's seat to provide a second set of eyes. We then started off into the cloud.

We quickly found that we were in icing cloud. Ice started to form as soon as we entered the cloud, but once we were committed there was no turning back, as the ice formation would make us too heavy. We had to go on and pray.

The descent was terrifying. Ice formed everywhere. Fortunately, most broke off. The cloud was so thick that we had no forward or downward vision other than about twelve feet around us. Aub's eyes were glued to his instruments.

The view from my seat was quite horrifying, just a grey nothingness all round us, and ice forming all over everything. It started to form on the propeller blades and I could see them start to slow with its weight. The engines coughed but, fortunately, didn't stop. Every now and then, a large block of ice would slip from the end of the propeller and crash against the fuselage or slide across the wing, but the engines continued on. This went on for minute after minute as we slowly descended. As we got closer to the ground, the cloud became denser and denser and I watched anxiously for hills, church spires, radio towers or the like.

We were getting desperate when I saw the ground. We emerged from the cloud base at 200 feet and, almost immediately, spotted a 'Drem'[3] system.

We turned towards it, but didn't know the station's callsign and couldn't raise it on the emergency frequency. Aub decided to land anyway, so we turned down the funnel and found the runway. It turned out to be Castle Donnington, a station similar to our own, and only a few miles from Church Broughton. We were made welcome, given beds for the night and drifted off to bed after Aub had reported to Operations.

The weather showed signs of lifting next morning, but we had to wait until after lunch before we could get away. The flight home took only twenty minutes.

An aircrew bus came to pick us up from dispersal. The WAAF driver pulled up, looked at our aircraft, and then, wide-eyed, gasped, 'J-Johnny! We thought you went for a Burton last night.'

Although Castle Donnington had reported our arrival to Group HQ, the latter had not advised our station and the Station Commander and CO had sat up into the wee hours waiting for a report that never came. We had been presumed lost.

We had been very lucky and, although we had followed careful and precise safety routines, the experience taught us that we should never have attempted the descent, but should have sought an emergency channel for instructions. Heaven knows what we would have done if Castle Donnington airfield had not shown up quickly. We could not have flown around in darkness at 200 feet for very long.

Nickels

The last exercise on an OTU was the so-called 'nickel'. This was an exercise which involved flying over enemy territory, usually with a load of propaganda leaflets to be dropped over an area of France.

Doug Glasson tells of his trip to Paris on a nickel:
My most difficult trip was the so-called nickel. When we did this trip we did not have Gee — it was not available on our OTU. Also, I had not been told that, if you had flu, you certainly did not take oxygen.

We set off for Paris and donned our oxygen masks at 13 000 feet. Very soon after, I was sick, to the point that I could not swallow the oxygen. It was an extraordinary experience. I had a dreadful vomiting attack and

had to get the bomb aimer to do my job.

By the time I had finally cleared out my stomach, we should have flown over our Paris turning point and been on our way home. To my horror, on returning to my desk, I found the bomb aimer sitting there completely frozen. He was as white as a ghost. We had been flying for an hour without a navigator.

All I could do was work from our estimated position of an hour ago and estimate our present position, assuming conditions had not changed.

For some reason, we had no radio — probably due to the weather conditions. The radio operator could not get a radio bearing for me. In fact, he was pulling the radio to pieces to try and fix it — he was doing a course in electronics through Cambridge, I think.

Our only chance of getting a fix was to use astro, but there was solid cloud above us. We tried hard to climb over it, but to no avail. We couldn't climb high enough.

I asked the pilot to fly a square search to try and find a break in the cloud. We had been attacked twice by fighters while over France and had suffered some damage. The evasive action had caused us to lose some height — we were down to 7-8000 feet — and we had to climb back up again.

We thought we were over the English coast and had started our square search when we were attacked, and badly hit, by flak.

We now knew we were not over the Channel, the cloud was still solid and we still had no radio. So we decided to fly west-north-west.

Bearing in mind the Golden Rule to be careful about the mountains in England, we flew at 4-5000 feet and started another square search. This was in the general direction of north to north-east. The cloud was still solid and the bomb aimer was told to get down and look for any break in it.

By this time, fuel was getting low and we faced a decision as to whether we assumed we were over the sea and ditched, or were we over land and jumped.

With petrol down to half an hour at most, the rear gunner yelled out 'I think I saw a light.'

'Right!' I said to the pilot, 'circle down and see if the rear gunner or bomb aimer can spot that light again. We'll hold on to that light, come what may.' We did find it again and followed it to Talbenny — the westernmost aerodrome in Wales.

---

3. Drem system — a lighting system for aerodromes, consisted of a ring of closely spaced lights in an oval pattern surrounding the aerodrome at a distance of a few hundred yards. Extra rows diverged from it and converged towards the end of the runway. Landing aircraft could easily follow the lights right onto the runway. The name 'Drem' comes from the name of the RAF station at which it was developed — Drem in Scotland.

We were delighted, when we landed, to find they spoke our language.

The weather was a most unusual pattern and had caused most of our problems. It was, however, the best sort of training for a navigator who had done all his training in Australia, in good clear conditions. You could look back and ask yourself, What should I have done?

On that score, we were interrogated at St Eval, by a squadron leader navigator who seemed to me to be a smart alec.

He was an Englishman and said, 'Well, Glasson, How can you explain yourself?' I said, 'I don't think there's any explaining to do, sir. I'm giving you the facts of what happened.' 'Well, what is your explanation for finishing up at St Eval?' I said, 'I suppose you are asking me what navigational error occurred, on my part, on this occasion. With your permission, sir, I would like to ask you a question. What would you have done in the circumstances?' He didn't answer the question.

Keith Burns tells of his experience while on his nickel in July 1944:

We had flown from 27 OTU Lichfield. Along with other OTU crews, we were to go to the Brittany coast, in Wellingtons, to release our load of bumph telling the French how best to hinder the enemy. The Normandy landing was four weeks ago.

On the way home, one of the Wimpies was coned over Jersey, where he had strayed off-course. We learned later that he crashed in southern England — there were no survivors.

We crossed the English coast at the appointed spot, and were proceeding northwards, towards base, when the rear gunner called the mid-upper, who was standing with his head in the astrodome — the Wellington had no mid-upper turret — and told him he was watching a twin-finned aircraft astern of us. He asked whether it was an Me-110.

They debated this for about a minute; when the mid-upper suddenly exclaimed, 'It's a bloody Lancaster and we are surrounded by them.'

The skipper quickly changed altitude to move out of the homeward bomber stream, which had been attacking flying bomb sites in northern France. We didn't want some-itchy fingered Lancaster air-gunner to think we were a Ju-88 as we had two engines and a single fin.

Harry Brabin describes his Nickel:

Our first flight out of England was a nickel — a leaflet-dropping mission — on Paris on 20 December 1943. Now we were to learn what Ops were really like — to see flak and feel it hit the aircraft like gravel hitting the underside of a car.

After the briefing at dusk, we donned our flying suits and fleecy-lined boots and were driven to the northern end of the aerodrome, where the Wellingtons were lined up. We were soon revving up for take-off and lifting off, one after the other. It was dark, and the stars were very bright.

When we reached the French coast, and again at Paris, we were greeted by a wall of flak. Hundreds of searchlights lit the sky and waved about.

We dropped our leaflets and headed back to the coast. Poor Sandy was in the rear turret without an electrically heated suit. It was near freezing and he was half-frozen when we arrived back — he had to be helped out of the turret. Although we lost several aircraft, the RAF did not count this trip as an operation.

There were many Australians among the instructors. These men, with their ribbons, their moustaches and slang, became our heroes. They had completed a tour of thirty bombing raids, which made them a superior race in our eyes. They all wanted to go back to their squadrons to do a second tour. We were quietly confident we would each do our best to uphold their standards.

Five aircraft had been lost on cross-country flights in a week. Some had vanished into the sea, one landed, unexpectedly, on top of a hill, bounced off, and slid down the other side without anyone being hurt.

Bas, our navigator, remarked, 'I wonder how many of this lot will see next year out?' I could only reply, 'It's hard to guess, but they say there are 5 per cent losses on each raid — and we are expected to do thirty ops.'

Arthur Doubleday had a narrow escape:

On the last training trip we carried photo flashes. They were about six feet long and six inches wide. They had a little propeller on them. We dropped them with our bombs so that our camera could photograph where our bombs hit.

The flash was carried in the flare chute. When the bombs were dropped, the radio operator had to count (to seven, I think) and then released the flash from the chute. The little propeller went round and fused the flash, which exploded about 5000 feet above the ground.

On this night, the flash didn't leave the chute until we landed, when it speared itself into the ground. The armament officer said, 'Ah, we'll find it in the morning.'

We did find it next day, and he said to the rear gunner and wireless operator 'Get a shovel and some disposal people and we'll dig it out.' I said, 'Look — let's just pull it out.' So we pulled it out and dumped it on the ground. The armament officer decided to light a cigarette — and up went the photo flash.

He said, a moment later, 'What are you doing to-day?' I said, 'I've got a cross-country at nine o'clock.' He said, 'Well, don't say anything about this.' But I couldn't help him because I had to report to Flight at once.

The flight commander said, 'What happened to you?' I told him. The Armament Officer was off the station in a few hours.

As it happened, I had got a hairline fracture of the jaw when the flash went up. But I was lucky — when a flash blew up at Rathmines one trainee was killed, one lost an arm and another was hospitalised.

Leisure Moments at OTU

Gerry Judd, at a Blenheim and Bisley OTU at Grantham, relaxed at Nottingham:

There was a dance on every Saturday night. My friend Harry and I met a couple of really good-looking girls there and were going out with them.

We were on A Flight and two of our fellows from the same course, Fred M. and John F., were on B Flight. They were off flying when we were on and vice-versa, so they went to Nottingham when we couldn't be there.

They met our girlfriends and told them that, unfortunately, we'd both been lost while flying over the Channel. It was no good waiting for us because we were gone.

It just so happened that a cold front came in over that weekend and flying was cancelled. Harry and I got on the train and went to Nottingham, arriving half way through the dance. It quite upset things for M. and F., quite ruined their chances.

Im Westen

Die Angriffe der Royal Air Force auf Deutschland sind nur der erste Abschnitt der britisch-amerikanischen Offensive. An der Vorbereitung des zweiten arbeiten England und Amerika Tag und Nacht.

Im Osten

Die ersten neun Monate des Russenkriegs haben Deutschland mehr Blut gekostet als die vier Jahre des ersten Weltkriegs. Wieviel Blut werden die zweiten neun Monate Deutschland kosten? Die ganze Welt ist sich darin einig: Deutschland geht seiner Niederlage unaufhaltsam entgegen.

Kann diese Schlächterei beendet werden, ehe es zu spät ist?

Noch einmal hat die britische Regierung in den letzten Reden von Churchill, Cripps und Eden den einzigen Weg hierzu klar und deutlich gezeigt:

1. DIE DEUTSCHEN müssen den Krieg, den Hitler um Jahre zu verlängern trachtet, selbst verkürzen.

2. DIE DEUTSCHEN müssen die Gangsterherrschaft, die Hitler und seine Spiessgesellen Deutschland auferlegt haben, selbst stürzen.

3. DIE DEUTSCHEN müssen sich in die Reihe der unterdrückten Völker stellen, die alle für die Wiedererrichtung ihres Staates kämpfen, eines Staates, der auf der Achtung vor den Rechten der Völker und des Einzelnen beruht.

Wenn Ihr glaubt, ein waffenloses Volk könne unmöglich gegen eine Besatzungsarmee und die Gestapo ankämpfen, dann lest die Geschichte auf der Rückseite: sie handelt von einem kleinen Volk.

G 27

Rear of nickel dropped by Arthur Doubleday.

Sieg ohne Waffen

EIN kleiner norwegischer Küstendampfer, die „Skjerstad", legte unlängst in Kirkenes am nördlichen Eismeer an. In ihm waren zusammengepfercht wie das Vieh 500 Männer, die zur Zwangsarbeit in ein Konzentrationslager abgeführt wurden.

Es waren norwegische Lehrer, an denen Quisling und Hitlers Reichskommissar Terboven ihre Rache kühlten, die Rache für eine schmähliche Niederlage in ihrem Kampf gegen 12 000 waffenlose Lehrer.

Schon im Herbst 1940 hatte Quisling im deutschen Auftrag versucht, Hitlers „Neue Ordnung" in die norwegische Erziehung einzuführen. Schon damals stiess er auf den hartnäckigen Widerstand der norwegischen Lehrer, Eltern und Kinder. So einmütig war dieser Widerstand, dass Quisling eine Anzahl seiner neuen Verordnungen zurückziehen musste. Während des ganzen Jahres 1940 blieben die norwegischen Schulen die Hochburg des nationalen Widerstands gegen die Verräter und die Unterdrücker.

Der Kampf verschärfte sich, als Quisling am 1. Februar 1942 norwegischer Ministerpräsident von Hitlers Gnaden wurde. Erst versuchte er es mit Drohungen: jeder Lehrer, so hiess es, der nicht der N.S. Lehrerfront beitrete, werde Stellung, Gehalt und Pensionsanspruch verlieren. Aber 12 000 von den 14 000 norwegischen Lehrern waren entschlossen, eher Stellung, Gehalt und Pension preiszugeben, als ihre Grundsätze zu verleugnen.

Und wie die Lehrer sich weigerten, der N.S. Lehrerfront beizutreten, so lehnten es Eltern und Kinder ab, sich dem Befehl Quislings zu beugen, der die ganze norwegische Jugend zwischen 10 und 18 in die nationalsozialistische Jugendbewegung pressen wollte. Die norwegische Geistlichkeit unterstützte und stärkte den Widerstand von Lehrern, Eltern und Kindern.

Als Quisling sah, dass alle Drohungen nichts nützten, griff er zur Gewalt. Hunderte von Lehrern wurden festgenommen. In Stadt und Dorf demonstrierte die Bevölkerung. Wo Polizei und Gestapo erschienen, um die Lehrer wegzuholen, folgten Eltern und Kinder ihnen mit Hochrufen und Dank- und Trostworten an den Bahnhof.

Ende März waren über 1 300 in Haft, grossenteils in den Konzentrationslagern von Grini und Brekstad, unter unmenschlichen Lebensbedingungen. Aber die norwegischen Lehrer begnügten sich nicht mit schweigendem Widerstand. Sie waren bereit, auch ohne Gehalt zu unterrichten, und als nach den Kälteferien die Schulen am 9. April, dem Jahrestag des deutschen Überfalls auf Norwegen, wieder aufgemacht wurden, verlas jeder Lehrer vor den Kindern eine Erklärung, in der es unter anderem hiess: „Ich werde von Euch niemals etwas verlangen, was ich als Unrecht ansehe, und ich werde Euch niemals etwas lehren, was nach meiner Ansicht nicht der Wahrheit entspricht."

Quisling musste einsehen, dass er machtlos war. So beschloss er, ein Exempel zu statuieren und zwar mit der Hilfe Terbovens und des Gestapohäuptlings Rediess. Hunderte von neuen Verhaftungen wurden vorgenommen, und der Terror erreichte seinen Höhepunkt in dem Abtransport von 500 Lehrern in der „Skjerstad" nach dem hohen Norden.

500 tapfere Männer müssen leiden. Aber weit mehr als 10 000 ihrer Lehrerkameraden führen den Kampf weiter, gestützt auf ihr Volk, getragen vom Glauben an Recht und Wahrheit. Ihr vorbildlicher Mut hat den Widerstandsgeist der Norweger so gestärkt, dass ihre Unterdrücker bisher nicht gewagt haben, es auf eine neue Kraftprobe ankommen zu lassen.

Front piece of nickel dropped by Arthur Doubleday for German troops.

Low Flying

Low flying was frowned on by the Command because of the unnecessary risks it entailed. Jack Stronach's tale is, therefore, unusual:

On one occasion, I was told to fly, with two other pilots, to carry out a low flying exercise. The aim was to give our AA gunners, and ourselves, practice in low-level attacks against gun sites.

Our aerodrome was a beautifully smooth paddock, or meadow, without runways. I dived down on the first gun site, reaching a speed of about 200 miles an hour, and going over the guns at about thirty feet.

The AA gunners just dropped to the ground!

I then came in really low. I followed the surface of the ground with the propellers more or less chewing the grass. Unfortunately, on the first run, there was a postman on a bike right at the other boundary. The poor devil fell off his bike as I skimmed over him.

I climbed up again and made low-level passes at the other gun sites. Eventually, I landed. Someone said, 'You're in trouble, Stronach. You've got to report to the CO.'

I went in to the CO. He said, 'Stronach, I'm going to fine you for dangerous flying. When you flew across the aerodrome, your aircraft disappeared out of sight.'

While the ground looked level, it actually had undulations and, as I crossed the aerodrome, I had flown a hollow and the aircraft had disappeared.

He said, 'Stronach, the fine is threepence.'

HEAVY CONVERSION UNIT

Having completed operational training, crews went to Heavy Conversion Units to get experience flying four engined aircraft. The majority of OTUs only operated twin-engined aircraft.

It was here that the flight engineer joined the crew. The course was brief in most bomber groups — only two weeks were involved. On 5 Group, HCU involved a conversion to Stirlings. This lasted a month and was followed by a four-day course of conversion to Lancasters. Then came the real thing — the Squadron.

Cross-Countries

One of the training exercises at heavy conversion unit was the flying of cross-countries. The hazards of these exercises are dealt with in the following incidents.

Keith Burns tells of an experience with Navy gunners:
Flying a Halifax II from 1652 HCU, on a night cross-country, we had left Fishgard on a course to Mull of Galloway on a moonless night. About ten minutes out, the air around us was filled with tracer shells.

The skipper called to the flight engineer to fire off the colours of the day, which he did, but the shells came even thicker. Maybe we had helped to target ourselves. The way out was a sudden change of course and a drop in altitude of about 10 000 feet in no time flat. We then completed the exercise.

Back at base, we reported our experience to the Intelligence bod. Rustling through a pile of papers, he asked, 'Didn't you get a convoy warning in that area?' 'No bloody fear!' We replied.

It taught us a lesson — those Navy gunners believed in the maxim -'ASK QUESTIONS LATER.'

Some pilots had difficulty with the conversion to four-engine planes. Alby Silverstone's pilot was one of these:
Our stay at Riccall was memorable, mainly because we survived to reach an operational squadron. Our flying experiences were unbelievable, right from the start Aub found the large four-engine aircraft difficult to control.

The Halifax I aircraft was powered by Rolls-Royce Merlin inline engines and control at ground level, for taxiing and straightening for take-offs and landings, was by use of the two outboard engines. To turn right, increase the speed of the port (left) outboard engine, and vice versa. Aub couldn't handle it at all, particularly for landings, he was too heavy-handed.

We left runways, we ground looped and we never knew what to expect from one landing to the next. The fact that we survived was only due to the sturdiness of the Halifax and a great deal of luck. Even some of our air trips were close to disaster.

The epic was a night cross-country around the Irish Sea. The exercise was to be flown at 16 000 feet and, as we climbed through the 10 000 feet mark, Aub ran through the oxygen drill and checked that everyone was connected. We climbed to our correct height.

We were proceeding smoothly when Aub interrupted and asked 'Hey! Are we upside down or something?' That startled us, but, being inexperienced, it didn't raise any undue alarm. When queried, he said that there were stars underneath us. These turned out to be the small blue navigation lights at the mastheads of a convoy of ships.

A short time later, Jim, who had an airspeed indicator at his navigation table, called out to Aub to find out why we were increasing our airspeed. When there was no answer, he called George, the flight engineer, who was stationed immediately behind the pilot, to find out what was happening.

George replied that we seemed to be descending and this was building up our airspeed. Despite all this talking, there was still no answer from our pilot. Jim picked it up at once. 'Check Aub's oxygen, George. See if he is connected.' That was it. He hadn't connected his own oxygen and had virtually passed out. He recovered immediately when he was connected. We settled down again from what could have been the finish of all of us.

Crashes

John Page had a couple of experiences at Heavy Conversion Units:
On one occasion I, as an instructor, was converting to instruct on Halifaxes. This involved flying in the right-hand seat — there were dual controls. A port tyre burst at speed on the runway. The aircraft lurched, the wheel dug in and the undercarriage collapsed. We almost cartwheeled but we all managed to walk away unhurt.

Just before my second tour I was taking off in a Halifax with a student. We were only at fifty feet when I noticed the rev counter on one of the starboard engines slip back. I knew the Halifax well by then so, although we were low, I said, 'I'll show you what a good aircraft this is.'

I feathered the engine and carried on with three engines.

When we landed and got back to dispersal we found the ground crew hadn't bolted the exhaust manifold on correctly. While we were taxiing it hung down, but when we took off it lifted up and the engine caught fire.

Three days later an inspection officer came up and said, 'We have to make a very thorough inspection of aircraft fires. Now you were in the cockpit. Tell me what happened.' I told him and he said, 'You knew it was on fire.' I said, 'No! I only feathered it to show what a good aircraft it was.' I don't think he ever believed me."

Noel Gilmour was lucky to survive a crash in early September 1945, when on a conversion course:
We were to become a Transport Squadron. We were not 466 Squadron any more — we were converting to Liberators. We did an hour and a half of circuits and bumps, followed by a late lunch with the instructor. He was an ex-Coastal Command type with 2 500 hours. Over lunch, he said, 'We'll do a single engine failure on take-off after lunch.'

The aircraft had been refuelled, so we took off again. As soon as the wheels left the ground, he cut the starboard outer engine. We climbed very slowly, preparatory to doing a three-engine landing. He said, 'We'll do some two-engine flying before we go in.' He cut the starboard inner engine. Both starboard engines were now out of action. The aircraft was going down — the ground was coming up — in the form of a hill.

There was not much control in a Liberator with two engines out on one side — it was difficult to keep it straight and level. So, we gradually descended towards this hill.

When it became apparent that something had to be done, the instructor unfeathered the starboard outer, which came up to speed (2500 revs). Then he turned the switches on, but nothing happened. So he unfeathered the starboard inner; and the same thing happened. We still had two engines out on one side and the ground was now approaching rapidly.

I looked across and there, in a row across the cockpit, were our pilot, the instructor and the flight engineer instructor. Leaning on the backs of the two pilots' seats were our second pilot (Frank Bax), our flight engineer (Roy Turner) and myself. There was a little Scotch Terrier (Bitser) at my feet. The wireless operator was in the 'attic' (the wireless operator's compartment back behind the wing), his name was Tex. Six people in two rows across the cockpit and Tex in his attic doing whatever he was doing.

Before we could find out why the two engines wouldn't fire, the aircraft stuttered, rolled over and slammed straight into the woods.

I remember, in the few seconds it took to reach the ground after it rolled (about ten seconds), thinking, 'Too bad. We're all going to be killed even though the war's over. I wonder what it's like to be dead.' Then there was a loud bang — like a man kicking on an old-fashioned kerosene tin. I didn't feel a bump or anything, but I came to sometime later on.

After a few seconds I realised I had been in an aeroplane crash and I was still alive. Then I could hear crackling and I knew it was burning. So I crawled my way out, burning my hands tearing hot metal away. I crawled on.

Our flight engineer called out for someone to help him — but I was beyond helping anyone at that stage. I thought both my legs were broken — but I found out later that was not the case.

The aircraft was very much smashed and scattered. The only part remaining was the tailplane and rudders. I believe it was scattered all over the place.

The crash happened late in the afternoon. The next thing I remember was being turned over and put on a stretcher. It was then night time. I was put in an ambulance — I kept on regaining consciousness and passing out.

I was conscious at an airfield — I think it was Tempsford — where I found myself lying on a table. Just as I woke, a medico was yanking my foot round in the right direction. That put me out again for a while.

Ultimately, they transferred me — about midnight — from Tempsford to Henley, where there was an Air Force hospital. I was put in a bed. According to a fellow inmate, I was swearing like a trooper. I'd never sworn before — I don't remember it.

I found out later that the crash had been seen and reported. But no stations seemed to be missing any aircraft, including our own station. After some double-checking, they found, about five o'clock, that one aircraft had not come back from circuits and bumps.

An ambulance was sent out, and it picked up the remnants. But when they added up the number of people recovered, they found only six. There had been seven on board — Gilmour was missing. The ambulance had to go all the way back to find me.

I'd crawled off into the woods. I didn't know where I'd gone. I'd crawled on my elbows to get away from being burnt. They found me because there was a dog barking. It was cuddled up to me. If I'd been left for the night I might not have made it.

I was in Henley, where I was patched up, for a couple of weeks. My hands gradually got better, the skin had been cut off where it was burned. There were great, huge, blisters.

Then I went to a great big peacetime hospital at Holton. I spent a couple of weeks there and was then posted to Brighton. The squadron had folded up.

Noel Gilmour returned to the site of the crash twenty years later. He interviewed a Mr Bennett, then eighty-one. He had seen the crash and told the story of what he saw: There were four of us fellows, my mate and two lads, coming back from the village after getting our weekly ration of butter and groceries. We saw the plane coming. It crashed in a paddock near us.

We raced to the scene and I called to the two lads to give me and my mate a hand. They wouldn't. They said, 'We're not going near it. It's probably got bombs on board.'

Anyway, we pulled one chap out. He had a fractured skull and quite severe burns [This was Ray Turner]. He probably would have pulled through one injury, but two proved too much. He died about 11 p.m. that night.

4

BOMBER COMMAND

3 September 1939 To 31 March 1944

BOMBING OPERATIONS

During the early war years, 1939-42, the men of Bomber Command paid the price of earlier government complacency and failure to provide funds to establish and maintain an effective air force.

It was true that a policy with regard to air operations had been developed. The role of the bomber was to be predominant. It was the bomber which would always carry the offensive to the enemy. Nothing could prevent this. Even flying in daylight, bombers would always get through flying above the flak and with gunners in the turrets fighting off any fighter attack.

This was the view of Viscount Trenchard, Chief of Air Staff from 1918 to 1929, and of his disciples who were to hold the key and influential RAF posts before and in the early years of the war. It was the view also of Lord Portal, who became Chief of Air Staff in October 1940 and was in that position for the remainder of the war.

As CAS, he continued the build up of Bomber Command as much as possible. In Portal's opinion no allied army would be able to land in Europe until German morale was broken. It was the bombers which would break that morale.

Despite its own conviction of the supremacy of air power, the RAF had not been able to prepare itself by the outbreak of war for the conflict in the air which was to follow. Indeed, all that Bomber Command had in 1939 was about 272 front-line aircraft consisting of Wellingtons, Whitleys and Hampdens. Its other aircraft, Blenheims and Fairey Battles, were not equal to the task confronting them.

Bomber Command's operations were virtually limited to leaflet raids and to mainly maritime targets.

On 12 April 1940, six Hampdens and three Wellingtons were lost in attacks on shipping in Stavanger. No further major daylight sorties were undertaken by Wellingtons and Hampdens.

The emphasis shifted to night bombing for major raids.

Sam Weller, then a sergeant pilot, flew in a sortie to Kiel during this latter phase in 1940 and he tells of the operational practices of the day compared with the more professional approach of 1944/45:

What I remember was the fact that we had no runways in those days. It was all grass, so there was mud everywhere.

If there were ops on that night, you did a test flight during the day. You went out to the aircraft and as second dickie [second pilot], you were last on, no seat for you, which was a bit off. Your job was a 'gofer' — go for this, go for that, do this, do that.

On this first trip I did — I had a Sidcot [flying] suit on, flying boots and all that. The Whitley being tail-down and the entry through the nose, it was very awkward to get in and as I had all this gear on, I had to put my arms up above my head and lever myself in. It was not easy with a parachute and all that jazz. I then sat there on the floor.

After the air test I went to briefing. Everybody knew where the target was during the day. There was no security at all. 'Kiel tonight,' and it was Kiel.

So, Kiel was the first place I went to. I had a sergeant pilot as captain. I remember how confusing the lights on the airfield were — no bright lights anywhere to guide you. You almost had to have second sight to find your way around an airfield at night in those days. Anyway, we took off and one of the jobs I had to do was to turn on the heater on the camera and take the cover off. I had also to load the gun in the front turret, which was very awkward to get into. I had to climb into it and put the seat down behind me. It was a Vickers gas-operated gun and I had these pans of ammunition which were stacked around the turret. I dropped one and released the spring and all the cartridges came flying out. I think they held about a thousand rounds and they were flying all over the thing — eventually, I got the gun loaded, it wasn't easy.

We climbed into cloud. The pilot said, 'Go down and put on the de-icers.' This meant my climbing through two tanks in the dark without my parachute. I got down there to the controls, worked the levers, and crawled all the way back. He then said 'It's okay now; go and turn them off again.'

This was at 15 000 feet. I had no oxygen, of course, and I was puffing my way down and puffing my way back all over the place. Later on, in Halifaxes they provided bottles [which could be carried or hung across the shoulders] — none of that luxury stuff then.

I remember that first trip well because the pilot kept on getting me to go and do this and do that and in the end it nearly did for me. In those days the door of the Whitley opened outwards and on one journey I trod on it and it opened, and I nearly fell out. We were over the North Sea at the time and I can remember looking down and gasping, 'My goodness'. The pilot was complaining bitterly about the draught coming in and I am hanging on to the door and trying to close it. I had a terrible time because of the air stream pulling it open. Eventually I got it closed. Anyway, they changed it later and made it open inwards, but not before it had scared ten years off my lifetime.

The other jobs I had to do were to supply the captain with cups of coffee every now and then. I didn't get a seat to sit on and all I did was be a spare wheel.

I remember another trip to Kiel. I remember looking out and I couldn't see anything. When we came back and went to debriefing the story the crew told made me feel, 'There's something wrong with my eyesight. There must be because I didn't see any of that.'

When we came out of debriefing I said I hadn't seen what they had described. 'Shut up and mind your own business!' was their response.

This exaggeration or line shooting — call it what you will — helped to give the wrong impression of the accuracy of bombing sorties.

In May and June 1940 the Battle of France required the diversion of the Blenheim aircraft of Bomber Command's No. 2 Group to the RAF's Advanced Air Striking Force. From bases in England they fought an abortive campaign in France.

During this period another turning point in policy was reached. Bomber Command was now allowed to operate east of the Ruhr, but was confined to purely industrial targets. Civilian casualties were to be avoided. This change took effect on 15 May 1940.

The bombing offensive was still relatively haphazard and was to remain so for almost a year.

The Battle of Britain in mid-to late-1940 called for a supreme fighter effort and the bomber activity was minimal. In addition, the Command had decreed that there should be a lull during the winter of 1940/41 so that operations remained virtually in limbo until February 1941. From March to July 1941 the Battle of the Atlantic required that Bomber Command resources be directed to attacking U-boat bases, U-boat construction sites and docks. It was not until July 1941 that the attack swung back to Germany.

During this period the maximum effort had been only 149 aircraft at any one time and missions were far from being sophisticated operations.

Targets were difficult to find. There were navigational problems due to the lack of effective navigational aids, and when found, the specific target had to be identified before attack. To say the least this produced problems.

Sam Weller tells of the difficulty of finding the target:
One of the troubles in the Whitley days was that I don't think we found the target most of the time. We had no way of knowing where the target was. Most of the navigators could use the sextant — they *could* use it but it was very rare for them to use it. I never flew with a navigator who used it in a Whitley. How could you know where you were without taking fixes on the stars when you had nothing else to guide you and often you relied on the forecast wind?

Jack Davenport operated late in 1941 and his experiences are pertinent to the task of target identification:
I went to Essen in late 1941 in a Hampden and I got into a lot of trouble. By the time I came to the target it was easy to identify and I was down fairly low having been forced down by the searchlights. I got my bombs right smack on the railway line and took what I felt would be wonderful photos. When I got back I was acclaimed because of the photos slapbang in the middle of the marshalling yards.

Later I was summoned down to Air Ministry and it was only when I got there that I identified the photo as being, in fact, the marshalling yards at Hamm. I'd been doing some twisting and turning before I got there and had ended up over the wrong target forty-five kilometres from Essen.

Jack Davenport tells of his sorties in 1941/42:
I flew initially on Hampdens. They were an old airplane at the beginning of the war but they were a very nice airplane to fly, though reported to have a problem with a stabiliser yaw. This, I think, was grossly exaggerated. They were extremely manoeuvrable so they were a good airplane in my view.

We started off bombing on what were referred to as 'activities on easy targets.' We then dropped mines, then leaflets and then bombed targets such as Hamburg and the Ruhr.

During this, the latter part of 1941, any visit to the Ruhr was a dicey experience. It was the industrial heart of Germany which was extremely well defended and they had hundreds and hundreds of searchlights. In those days a raid of 100 or 200 planes was a fairly reasonable turnout.

We adopted the policy of going in lower because we thought it was safer. We'd go in at about 7000 feet and maybe lower than the main run of squadrons. We thought that if the German ack-ack fired at us then the shell on the way up, before it exploded, actually had to hit us. However, if we were on top when the shells exploded there'd be quite a bit of shrapnel around.

We were given a route to the target, but we didn't have the stream operation which later in the war became the norm.

I got hit most on a trip to Hamburg and on that occasion I collected, I think, about 177 holes, but I got back without too much trouble. I lost an engine on the way back too, but that wasn't a problem. I was apprehensive but the remaining engine worked all right.

Night fighters were always scary because you normally didn't see them.

It was the devil if you got caught by searchlights. One got you and the whole cone of searchlights would come up on you and they lit up an area of some acres. We developed a couple of techniques which sometimes helped.

We knew that some of the searchlights were sound controlled and we would take along a couple of empty beer bottles and drop them over the side and on one occasion we saw the searchlights follow the beer bottles down — that was encouraging.

I was only really caught badly in the searchlights on one occasion and then the beer bottles didn't work. I took fairly vicious evasive action because I was apprehensive of the night fighters. The problem was accentuated by the fact that the light from the searchlights was very very blinding and if you looked out it was very difficult to read the instruments — but the night fighters in those days weren't as effective as I'm sure they were later on in the war.

Our navigator had a very difficult task because frequently the aids that we had wouldn't work very well. The old d/f loop[1] was something you could spend twenty minutes on without success and he needed that twenty minutes for other things. So radio fixes were a lot of effort even though attempts were made to make

them more reliable. Dead reckoning was relied on and, of course, map-reading.

The Germans were very clever. They provided false targets which were five or ten miles from the true target and then lit the false target with artificial fires and things of that nature. At Hamburg, there was a lake which should have been clearly identifiable on most moonlight nights but they covered it with wooden rafts so we couldn't see the water.

We didn't have TI's[2] but our attacks weren't as bad as it sounds and though much has been written about the inaccuracy of Bomber Command, I think in these early days there were some very great successes, e.g. the Hamburg raids. They were very successful.

Sam Weller flew Whitley aircraft in this phase and had the misfortune to land in the sea. He ditched and lived to tell the tale:

I went to Stettin in a Whitley and we ran out of fuel on the way back and went into the water. We all survived.

We were airborne for eleven hours and five minutes before we ditched. We told base we were going to ditch per our radio.

It was four in the morning. When we first hit the water, the tanks being empty and gear up[3], both engines cut within a few seconds of one another. We had only

1. D/F loop — a circular aerial which could be rotated to measure the bearing of the aircraft from a radio station and so determine the aircraft's position.

2. TIs — target indicators. These were flares of various colours which were dropped on the target by special target-finding aircraft as aiming points for the bombing aircraft

Sam Weller as a sergeant on 102 Squadron RAF, Topcliffe, Yorkshire, planning his route to the target.

been able to maintain a height of 4 000 feet until the time of ditching. It was pitch black and we had no idea what the waves were like, we couldn't see them. Luckily for us the sea was calm but there was a slight swell.

The thing was to just let it fly down. When we hit, it bounced, it hit a second time, a much harder hit than the first time. The force smashed the perspex and water came rushing into the front of the aircraft and there was a great deal of noise. The plane floated because the tanks were empty and there was considerable buoyancy. We all got out.

Getting out of the aircraft in the dark — I didn't know of the drama that went on. In those days the dinghy was in a sort of valise (a dinghy pack in the aircraft) and not in the wing like a Halifax. It had a cord on it and it was the rear gunner's job to throw it out.

He told us he threw it out and the cord broke, but what happened I think is that he let it go. Fortunately he jumped out afterwards and got it attached.

The dinghy floated but unfortunately upside down because, in those days it had the bottle[4] on top and it acted as a keel, and it was upside down. There was no way we could turn it over. So we got in it upside down. By the time I got there they were all sitting in it and nobody was going to get out.

We were picked up about five hours later. We could hear the Air Sea Rescue launches, but although we could hear their motors they could not hear us.

We had a kit which was wrapped in canvas. When it was wet you could not open it because your hands were too cold. To help open it you had a great big jack knife which again you could not open. Later in the war things changed but at this stage there was no way we could get that kit open. Somebody said there was rum in it. If there was rum in there it stayed there — we certainly didn't get a drink.

Eventually we were picked up by a fishing smack, accidentally. As we went back to shore the Air Sea Rescue launch came alongside and said, 'We'll take them off you'.

The reply was, 'I'll keep them — you take the dinghy'. Anyway they took the dinghy and we went to Bridlington Harbour. The tide was out so we had to climb a little ladder and go from boat to boat — almost more dangerous than being picked up at sea.

They took us to hospital, dried us out, and then sent us back to Topcliffe. The squadron sent an old truck over to pick us up. It was about fifty miles from Bridlington to Topcliffe across the wold on a cool day. When I got back there I had lost my flying boots (they had filled with water and disappeared). I was in bare feet and I'd also lost my hat.

[3] Gear up — the undercarriage was retracted. If the undercarriage was down when the aircraft touched the water it could flip the machine on to its back.

[4] Bottle — a small cylinder of compressed nitrogen which automatically inflated the dinghy when it struck the water.

Air Sea Rescue launch, North Sea.

The group captain wanted to see us and I made some statement or other. When I came out the station warrant officer was there and I stood before him, with no boots and no hat. He wasn't worried a bit about the boots and me standing there in the cold. 'Where's your cap? You're a troublemaker, the group captain is always complaining about you.'

I knew that was a lie because I'd just been talking to him. 'Where's your cap?' So he got his book out.

'My cap is at the bottom of the North Sea. If you want it you can go and get it.'

The general pattern of an attack in this period is explained by Arthur Doubleday:
The attacks were almost haphazard in my view.

In the main, squadrons were left to their own ideas on the attack. The plan of attack was in the hands of the squadron commanders. The time you took off and all the other details were fixed by the squadron commander who was assisted by advice from his flight commanders. Group Headquarters did not enter into it. This seemed rather odd; if there was an ENSA concert on some commanders would decide to take-off after the concert was over. Some took off quite early, so the raid could be spread over a few hours. There were a lot of individual runs.

Change came in 1942 with the appointment of Air Marshal Arthur Harris as AOC-in-C Bomber Command.

Much criticism has been levelled at Harris by postwar historians and politicians, but there is no doubt that he made Bomber Command a strong force with sharp teeth. He was dedicated in his purpose and a man convinced of the strength of his cause. He later said, 'When your grand-children ask of you what you did in the Great War, you can say — I won the war.'

A man of purpose, he waged a campaign from February 1942 to the end of the war. There were three main phases of Harris's campaign: 22 February 1942 to 1 February 1943; 2 February 1943 to 30/31 March 1944; and 1st April 1944 to 9th May 1945 (Covered in Chapter 5).

22 FEBRUARY 1942 TO 1 FEBRUARY 1943

New navigational equipment and new planes were becoming operational as Harris assumed control.

Halifax aircraft, first operational in July 1941, were now arriving in large numbers. The powerful Lan-caster flew on operations for the first time on 11 March 1942.

At approximately the same time the navigational aid 'G' was available for navigators. This device had an effective range as far as the Ruhr and provided an electronic means of calculating one's position more accurately than previously. It was hoped that it would allow greater concentration of planes over a target and that this could be done in a shorter space of time.

Attacks were made against the ports of Rostock and Warnemunde during April/May with between 100 and 200 bombers taking part.

On 8/9 May 1942, 193 bombers were sent to Warnemunde, on the Baltic Sea. The Heinkel factories here were building Me 109s and Me 110s destined for use on the Russian front. The bombing was not effective — factories were difficult to find and to hit, given the inadequacies of navigational and bomb aiming equipment at that time.

Arthur Doubleday flew a Wellington from Australian Squadron 460 on this attack:
I ran into searchlights and was hit by flak as I came off the target. The oil pressure in the starboard engine fell to zero. It had been hit in the rocker box. The engines were radials and had these rocker boxes above each cylinder. I said to the gunner, 'I want you to pump five strokes per minute from the auxiliary tank. Now ten per minute.' Up came the oil so we could reduce to three or four strokes per minute — then pumped three per minute and so on. Eventually he pumped sixteen gallons of oil in this way. There was some anxiety about that.

Harris now switched to a major raid on Germany on a scale of unprecedented strength. The great raid of 1000 bombers was planned for 27 May or as soon as possible thereafter, depending on the weather. The target intended was either Hamburg or Cologne.

Sam Weller and Arthur Doubleday tell of their participation. First, Sam Weller:
I had done a few trips on Halifaxes and I had a crew. Then they started to deliver squadron planes. I was presented with a brand new Halifax — W1066 G, George. It was a Mark II with Merlin high-powered engines. We did the synchronising of the guns and all that.

Suddenly the atmosphere changed. There was something going on; we didn't know what. There was a sort of electricity in the air. There was no leave to be taken and those already on leave were recalled. This was always a bad sign.

So, here's my 1066 — brand new. I had done a few flights. We had a new wing commander and he sent someone to tell me there were ops on that night and he

Result of fighter attack on Wellington, F for Freddie, of 460 Squadron RAAF, Breighton, Yorkshire.

was going to take my aircraft I said, 'That means I am not going to operate.' He said, 'Oh yes you will be. There's another aircraft coming for you.'

'Where from?' I asked.

'From Marston Moor.'

'You mean it's a training aircraft?'

'Yes,' he replied, 'but it's a good one.'

'How do you know?'

Anyway, this poor old thing arrived. It was a training Mark I! It had no front turret, no mid-upper, only a rear turret. There were no modern appliances in it at all. It had been a training plane and its wings sagged and it had the original Merlin engines.

The operation was definitely ON. We didn't get to briefing however. It was cancelled on about 28 May. It was cancelled again the next day.

On 30 May Bomber Command Headquarters was becoming worried. The attack needed to be mounted with a full moon and the moon would now start to wane. All these aircraft had been assembled and nothing was happening.

So, on 30 May it was definitely ON and we went to briefing. There were to be 1042 airplanes on the raid. You can imagine the reaction because we had usually operated with about 200 at most. An enormous roar went up.

Four Group was to be last on the target. First was 5 Group and I think Three was next, then 4 Group.

At that time of the year in Yorkshire it didn't get dark until late.

When we went out to the aircraft, just as I arrived all the bombs fell off. So time was lost while the armourers put them back. I was late taking off and got on target about two-thirty in the morning and I could see the dawn coming up.

My plane was so slow it got up to about 6 000 feet and that was about as far as I could go. It was very very slow. When I got back to debriefing everybody was complaining about the icing. I said, 'What icing? I didn't see any icing.'

The target was well lit up before we got there. It didn't need much navigation — you could see it for miles.

By the time we arrived I think the ack-ack were tired — there was only a bit of activity.

Arthur Doubleday:

The number of planes in the raid on Cologne was misleading as to Bomber Command's strength. Not more than half were from the regular squadrons. The rest came from OTUs and other sources. Some of the aircraft were not properly fitted out for carrying bombloads. There were even some Hudsons, I think.

It was a Saturday night. It was very good weather. Some reports say it was bad weather on take-off but that is not my recollection at all.

The nearest cloud that I could see was almost as far away as Colombo and as we went over the North Sea there was the biggest harvest moon. Yellow and big and bright and you could see aircraft half a mile away. It was a magnificent evening. We weren't in the first so-called waves that went over but we saw the target from a fair way back. This was the first time I hadn't been picked up by searchlights and ack-ack.

The whole searchlight belt had been shifted and placed around the city. We were delighted with this change because we didn't have flak until we got into the target whereas, on previous sorties, we had it for fifteen minutes going in. They couldn't man the whole belt. It was bad for fifteen minutes but we weren't holed on that raid.

After I had bombed I went over and circled Bonn — just to have a look — about twenty miles away and there were no defences there. Then, because it was a moonlit night and there was nothing higher than the power lines, we just went down on the deck and came back at about 400 feet.

John Page trained on Oxfords and then went to a Hampden OTU On 25/26 June, while still at OTU, he went on a thousand-bomber raid to Bremen, inexperienced as he was:

At that stage, I had a total of about 160 hours flying experience — 26 on Hampdens. The crew consisted of a pilot, wireless operator, top gunner, under gunner and a bomb aimer who sat in the nose and was also the navigator.

The Pathfinder Force

This was formed on 15 August 1942. It was ultimately an elite force formed to provide more efficient marking of targets. Ted Eagleton gives a brief description of its method of operation:

Pathfinders were normally crews that were selected by squadrons for their ability and were sent off to lead the way for the main force.

They had what they called skymarkers and groundmarkers — red, green and yellow. They used to drop these to mark turning points on the way to the target. They would also then mark the target.

We used to have Master Bombers who would go down low and call the main body of Pathfinder boys in and tell then where to drop their markers so that we could bomb on them.

The Master Bomber was a hazardous operation and Guy Gibson, the Dambuster wing commander who won the VC, was one of them. The last they heard of him was him down low calling his planes in and saying, 'Come on fellows, It's easy tonight.' Never heard anything more of him. [Gibson crashed into a hill in Holland 60 miles from Rheydt and is buried there.] Leonard Cheshire, VC was also one of the top Pathfinder boys. Those guys didn't seem to have any fear at all.

Damage to Arthur Doubleday's Lancaster — turret and gunner have disappeared.

Italian Raids

Late in 1942 a number of raids were mounted on Italian targets. This required a long trip of about ten hours with extreme cold over the Alps. The targets were not as hotly defended as those in Germany. Bob Henderson and Sam Weller tell of their experience.

Bob Henderson:

I went to Turin twice. One trip was ten and three-quarter hours — a long time to be in control of an aircraft. I was in charge for all of the time except for a very short time when I had to get out and relieve myself over the sea.

Because I urinated in the porthole in the bomb bay I might as well have done it in my pants. When I finished I had a square patch the same size as the port saturated with urine.

We went directly down to Turin and having bombed there went down towards the Mediterranean to Biarritz on the border of France and Spain at zero feet and then we had to continue on about 400 miles due west and turn north east to Portreath in Cornwall.

It was a most interesting trip.

Sam Weller:

I remember the Alps. The trip to Italy from the UK was nine or ten hours. As I came to the Alps for the first time I saw them in the moonlight. I seemed to be flying into them. I was at 18 000 feet and I said to the navigator, 'What height did you say the highest mountain was?' He said, '15 000 feet.' I said, 'Are you sure? Come and look.'

We seemed to be flying into them in the moonlight. It was an optical illusion, really, and we went over the top all right.

There was no opposition. The main problem was icing.

John Page has this to say.

I went to Italy in a Halifax with Merlin engines. I was on 78 Squadron at Linton-on-Ouse. On the same station was 76 Squadron commanded by Wing Commander Leonard Cheshire.

The Halifax was a very nice aircraft — I had no complaints about it, although it had a bit of a coolant problem. On a trip to Turin in Italy I lost an engine due to this and had to return early. I had done seven hours flying on the trip so we were down as far as the Alps before returning. Although the Halifax II performed quite well, it didn't perform well without one engine and we wouldn't have cleared the Alps.

I did another trip to Turin (20 November '42), which was about ten and a half hours, and we got our bombs right on target.

We had George, the automatic pilot, which we could use for a time, but if we were over enemy territory with George engaged we kept our hands on the controls so we could take quick evasive action if necessary. We weaved, sometimes 10 deg. off course for five to ten minutes and then turned back on course.

2 FEBRUARY 1943 TO 30/31 MARCH 1944

During this period there were three major battles: The Battle of the Ruhr — 5/6 March 1943 to 24 July 1943; The Battle of Hamburg — 24/25 July 1943 to 3 August 1943; and The Battle of Berlin — 18/19 November 1943 to 31 March 1944.

Sam Weller gives us a brief introduction to the period. He had now finished with Wellingtons and was entering the Halifax segment of his career:

1943 was a bad year — we had a bad time. Halifaxes were still operating with Merlin engines and were down in power. The Halifaxes were getting knocked down more than the Lancs. They were always below. In theory we were supposed to get up to 20 000 feet but we only got to 18 000 because the plane couldn't make it.

They took the Merlins away and came up with the new Halifax Mark III with Bristol Hercules engines and that made a big difference. They had something like 1600 hp as opposed to 1200 hp. They also changed the shape of the wing and reconstructed the shape of the tail.

At the beginning of 1943 we went to Berlin and Leipzig frequently.

The year saw some disasters too when the winds all changed and everybody got blown off-course and blown away. We had Gee[5] — and H2S[6] and all that stuff — and stuff we didn't know about — mike in the engine which jammed frequencies [of German fighter controllers].

The Battle of the Ruhr — 5/6 March 1943 to 24 July 1943

The industrial region of the Rhine and the Ruhr was the greatest in Germany. The great Ruhr coalfields ensured a ready supply of coal for iron and steel production and hence munitions. Large industrial centres were sited at Essen, Dortmund, Dusseldorf, Duisberg and between Wuppertal and Huyen.

5. Gee was a piece of equipment which allowed the navigator to find his position accurately. He measured the time taken between the arrival of short wave pulses put out from a chain of ground stations.

6. H2S — a method of radar map reading used as an aid to navigation and bombing obscured targets.

There was also much other industry important to the German war effort. Transport for delivery of products was readily available. The Rhine was navigable for ocean- going craft as far as Duisberg. There was a network of roads and railways.

Here was a vital centre reasonably close to England providing targets for the RAF.

By now, navigation had improved with the development of new equipment and techniques. Gee enabled navigators to find their position, and Oboe[7] could be used to achieve target marking accuracy.

Four-engined bombers made up 70 per cent of Bomber Command's operational force. Between 500 and 800 planes could be put into the air on regular raids.

So the Battle of the Ruhr began. It lasted five months, during which 870 British aircraft were lost and 6000 aircrew were listed as missing.

Doug Glasson was engaged in the nine sorties to the Ruhr between 26 April and 28 June 1943. He comments:

The Ruhr — 'Happy Valley' — was at that time the area for attack by Wellingtons. It meant taking a 4 000 lb bombload on a sortie of between three and five hours. From that point of view it was a comfortable operational area for Wellingtons.

Because we were a somewhat slower aircraft we were usually placed in the waves up towards the front. By the time we got out of enemy territory we were about two-thirds of the way back in the stream. From April 1943 to July/August 1943 the Ruhr was our favourite area for attack.

The Ruhr was the hot spot. The Germans were 90 per cent certain that the Ruhr would be the target — they virtually knew we were coming. When you were taking off they had good knowledge that you were operating. For a start there was a lot of activity during the day — air testing, etc. and their intelligence was such that they would be well aware that there was something on that night. Over that period if there was a major operation in the wind they would concentrate their fighter and ground forces for the probability of a Ruhr attack.

Flak and searchlights were concentrated as the defences overlapped in a number of areas.

Peter Balderston:

On the night of 28/29 May 1943 we got off after four prior briefings bound for Wuppertal. Our Gee was useless but we didn't scrub. We decided that if we hit a pinpoint on the enemy coast we would be able to keep going. With the navigation equipment useless we could have turned back but we wanted to get it over and done with.

So we continued on, crossed the enemy coast and we were within half a mile of our proper course. We flew on from there. We didn't know where we were

until we saw the flak coming up and then a great burst of flame came round us and the pilot threw the aircraft around so much he turned it upside down. That toppled the artificial horizon and the other instruments. We then decided we were at Cologne instead of Wuppertal, three or four miles away. We were alone over Cologne and the target of their whole flak defence.

We eased out of that and got over to Wuppertal and dropped our bombs.

We headed home on what we thought was the right course. It was normally only about an hour from the enemy coast to the Yorkshire coast. We kept flying; the hour went past and then the hour and a half and we still were not in the area of base. We decided then we must be flying north up the North Sea. We turned west and hoped we hadn't gone past the north of Scotland but saw the coast in a few minutes.

By this time we were running short of petrol. So I started calling out on my R/T, 'Hello Darky.' This was a recognised distress call. Immediately an aerodrome lit up right in front of us. We were right on the approach and we just came in and trundled down the runway. A fellow with a blue light led us to a parking bay. We were just getting into the bay when the engines cut. We were down.

Bob Henderson had an eventful trip to the Ruhr:

The target was Oberhausen on 14 June 1943. There were 197 Lancs on the target. We lost seventeen.

On the climb before setting course to the target I noticed that the rear gunner was getting talkative and becoming quite irrational. I was a bit frightened he would swing his turret round and affect the trim of the aircraft His incessant chatter was driving us all mad.

By this time it was completely dark and I told the flight engineer he had better get an oxygen bottle out, take the wireless operator with him, and see what he could do about getting the gunner out of the turret. There was obviously something wrong with the gunner's oxygen supply.

It was an extremely good crew and these two fellows went down over the great spar and down to the turret. They had to get it centralised, open the door, get the gunner out and take him to the rest position.

By this time the gunner was in no mood to give any assistance and was very difficult, He was nearly unconscious and stayed in the rest position for the remainder of the trip.

---

7. Oboe — was a radio aid used to facilitate bombing through cloud, Which allowed aircraft to bomb targets as small as a factory or a military installation. It was used mainly as a target - marking aid and to allow flares to be dropped accurately. Oboe had limitations in that it required a consistent course for ten minutes before reaching the target area. It was used extensively by Mosquito target-marking aircraft

None of us was very happy about doing a trip without a rear gunner but we still had the mid-upper so we pressed on and did the job without any argument.

Had I been on a second tour I might have referred to the crew for a consensus but on the first tour I felt I was the boss and it was my responsibility to make the decision.

Bob Henderson went to the Ruhr again on 22 June 1943 when the target was Mulheim. There were 557 planes on this sortie — thirty five were lost; that is, 6 per cent.

We were coned [caught in the beams of a number of searchlights] coming out of the target and coned again a little later on. While we were over the target with our bomb doors open we were hit by flak which knocked out the port outer engine and the hydraulic line. This meant there was no way in the world we could get the bomb doors closed.

We lost height fairly rapidly down to about 4 000 feet and by this time the searchlights found it difficult to pass us from one to another so we slipped them and managed to make a bit of height.

We were flying along with the bomb doors open — which is almost as good as flying with full flap. Speed was dramatically reduced due to this combined with the loss of the port outer. We began to lag seriously behind the main stream — in fact we were given up by base who were surprised when we called up.

How did I get down? I had no hydraulics so what was the best thing to do? A belly landing or whatever? However I was able to use an emergency high-pressure air bottle in the hydraulic system which allowed us to make a more or less normal landing.

I decided to get the wheels down and they came down. The next thing was to put a bit of flap on but the bomb doors were still open. I got enough flap down and then, with what was left in the system got the bomb doors closed. That done, I could concentrate on landing without the port outer operating — which was hard enough.

Keith Skidmore was shot down on the last attack of the Battle of the Ruhr. The target was Remscheid on the night of 30/31 July 1943. This was the true end of the Battle of the Ruhr, although it is shown in most accounts to end on 24 July.

My duty, as mid-upper gunner, was to vacate my turret at a certain distance from the target and disperse tinfoil (window) out of the flare chute in the fuselage. This I did.

On reaching the target we dropped our bombs and, to my relief, we were on our way home. Suddenly we were coned in a barrage of searchlights. The immediate conversation went something like:

Gunner to Pilot: 'Ron! What will I do?'
Pilot to Gunner: 'Stay where you are!'

A few seconds or minutes later a blue searchlight (radar controlled, I believe) got our plane. All of a sudden! Flash! Bang! The mid-upper turret was a shambles. The rear hatch door was twisted and warped, the intercom gone and my oxygen mask failed.

The next thing I knew a fighter attacked. I think I was shaking so much that the bullets couldn't hit me! I think the fighter made three attacks — it could have been two. The bullets ripped right along the fuselage. I did see one of the crew hit. By this time the front of the plane was all alight.

What could I do? I put my parachute on[8] but there was no way I could get out. I couldn't open the entrance hatch, the flames and heat were unbearable. All of a sudden I saw the rear gunner (Bluey Rourke) get out of his turret to get his parachute. He spoke to me but without intercom I only sensed he was baling out.

'Gee! I'll follow him!' He got into his turret and turned it to the beam. I waited a short while, then tried the manual lever to wind the turret back. It didn't work. The turret was jammed. Bluey's legs may have been caught or trapped while trying to bale out. I'll never know. He was killed.

The most amazing thing happened to me when I realised there was no way out. I became quite calm, no longer scared. My Mum and a bloke I believed to be Jesus Christ came into the plane. Each took one of my hands and they walked me down to the centre of the plane and told me to lie down, which I did. Then I passed out.

I have no idea what happened after that. The next thing I knew I was falling through the sky. 'Good Heavens! I'm out!' Then I suddenly realised I couldn't see. 'God — I'm blind!' I rubbed my eyes and I could see. The heat had scorched my face and my eyelids had stuck together. What a relief!

I felt my chest for my parachute. It wasn't there. 'I'm gone again!' But, as I looked upward, I saw the parachute bag above my head. I reached up and pulled the ripcord. The parachute opened.

8. The pilot of a heavy bomber sat on his parachute which was permanently attached to his body. All other crew members wore a harness to which a separate parachute pack could be clipped in an emergency. The packs were stowed adjacent to the crew members' work stations. The turrets were too cramped to accommodate the packs so the gunners had to get out of their turrets to get their packs and clip them on before baling out.

The rear gunner could re-enter his turret, leaving the doors open behind him, turn it to one side and fall out backwards. In the later stages of the war the rear gunner was equipped with a pilot-type parachute so he did not have to leave the turret to get his chute.

The mid-upper gunner had to leave through any escape exit he could reach — generally the rear door but this was inoperable in this case. In general the whole crew, except the rear gunner, tried to leave through the front hatch. In this way the pilot knew when the last man had gone and could himself leave.

The free fall prior to this is the most amazing feeling I have ever experienced. The opening of the parachute pulled me up with a bit of a jolt. I'm glad it opened. I looked around and could see searchlights.

I drifted through the clouds and I thought I was going through more clouds when I hit the ground. It must have been a ground haze. What a landing!

The Battle of Hamburg — 24/25 July to 3 August 1943

Hamburg was the most important of the German ports. For RAF operations it had physical features which were readily recognisable for map reading both by eye and by H2S ground returns. Moreover, attacks could be made from the sea, thus avoiding the dangers of flying all the way over enemy occupied territory with the attendant dangers of flak concentrations and enemy fighter bases.

In July the attacks commenced in earnest. A new anti-radar device — 'window'[9] — was launched on the attack of 24/25 July to confuse the radar control of flak, fighters and searchlights. As a result, the German defences were completely disrupted. This attack was backed up by another three — on 27 July, 29/30 July and 2/3 August.

On the first three occasions 700 aircraft attacked. Bomber Command's final attack of this series failed. Because of bad weather only 425 out of 740 aircraft attacked the target.

A new diving tactic had been introduced and an acquaintance of Sam Weller had just dived to a second level to throw off the searchlights and fighters. He relaxed. . . Sam Weller explains:
He came out of the dive and put George [automatic pilot] in because he wanted a cigarette — he was a heavy smoker. He got out a cigarette and was just lighting it when he was bounced by a fighter.

The rear gunner and the mid-upper were screaming their heads off so he pulled the auto pilot out and pulled the plane over with tremendous force, which pushed him out of his seat. He didn't know what followed at the time but reconstructed it later.

It seems he hit the feathering button[10] with his head. One engine was already stopped so now he had two engines out on one side and right over and doing a 360 degree turn at a very acute angle. After much buffeting around he was down to 2 000 feet. By this time he had lost the fighter.

While they were at debriefing he told them the dramatic story of having lost an engine and having another shot out by a fighter. He didn't tell them about the cigarette.

Afterwards, the station engineering officer came and said, 'Nothing wrong with that engine.' He said, 'There must be.'

'There's not a mark on the plane.'

Someone asked him, 'What's that mark on your head?' [It was where he had bumped the feather switch.] He thought about this, realised what had happened and shut up.

Attack by fighter was responsible for more losses in Bomber Command than any other factor. Initial heavy losses in 1939-41 caused the RAF to switch to night bombing as darkness would protect the bombers to some degree. However, they still had to fly approximately 200 miles before attacking their targets. All this was over enemy-occupied territory.

By 1942 the Germans had developed a defensive chain of radar stations. The Kammhuber Line was developed to counter the British bomber offensive. It consisted of ground radar stations positioned twenty miles apart. The line ran from the north to the south, crossing Denmark, Germany, Holland, Belgium, and eastern France.

This enabled early warning to be given by Freya long-range radar stations to fighter stations equipped with short-range Wurzburg radar by which the fighters were directed until their airborne radar detected the enemy and helped the aircrews to make visual contact and attack the bombers.

The German defence was very efficient and inflicted considerable losses on the attackers.

As mentioned above, on 24/25 July 1943 the British used their anti-radar countermeasure — window — on a raid on Hamburg. These strips of foil blotted out the German radar screens, confused their radar operators and were highly effective.

Ted Eagleton gives an account of the four raids:
These were the most interesting of the raids I took part in — mainly because Bomber Command was trying out a new radio device which was called window — metal strips thrown out of the aircraft to upset the German radar and the German night fighters.

Fortunately for us the first night was a clear night and very heavily defended by numerous searchlights. The effect of window was devastating for the Germans. I'd never seen anything like it. The searchlights were just wandering aimlessly around the sky. They didn't cone anybody.

Being coned over a target was rather a horrifying type of thing. You either got away by diving or by doing a

---

9. Window — strips of metal-coated paper released from aircraft Falling slowly through the air, they caused radar echoes which effectively jammed the German radar screens and drastically interfered with the control of the defending night fighters.

10. There was a 'feathering button' for each engine, which was only to be used when the engine was being stopped. When the button was pressed the blades of the propeller on the corresponding engine were turned so that they were parallel to the propeller axis. In this position they offered least air resistance. This could be vital in a damaged aircraft with one or more engines out of action.

corkscrew and that was always dangerous because you could collide with another aircraft Corkscrewing was a manoeuvre to try to get away from German fighters or get yourself out of a searchlight situation. You turned hard aport and lost 300 or 400 feet and then you turned starboard and flew straight and level provided you had shaken off the enemy.

Hamburg was one of the most heavily defended targets in Germany at that stage and it was quite frightening.

I was doing a course when the squadron was first briefed for Hamburg. The first raid was cancelled and I arrived back from the course and sure enough, one crew went on leave and I went that night on ops. There were four raids and I took part in all of them. They turned out to be reasonably easy trips as a result of the introduction of window.

However, the fourth was a shocker. We had bad weather forecast and when we were getting close to the target we ran into cu-nimb[11] clouds. So much so that the lightning was running up and down the guns. My rear gunner, Barney McCosker, came up on the mike and said 'Hey Ted! If this lightning doesn't stop playing round the turret I'm getting out. There's not room for two of us in here.' Having got through that bank of cloud we came into a clear patch on the turning point to go into the target

Giant Wurtzburg a German early-warning device which located enemy bombers.

and, of course, that's where the fighter boys were hovering round.

Fortunately for me, slapbang into another cu-nimb we went again, so we really bombed blind on Hamburg.

We were attacked again by a fighter coming out of the target but by corkscrewing I managed to throw the fighter off and returned to base with a clear run. I was flying a Wellington.

Wellingtons of 466 Squadron RAAF, Leconfield, Yorkshire, in flight.

Doug Glasson also had an interesting weather experience on the raid of 2/3 August 1943:

I looked out and it looked like an enormous black canyon that we were flying into. There seemed to be little point in losing height. On the other hand we were going to dive into the mess. The St Elmo's fire was quite incredible.

The next thing the pilot yelled out, 'We're losing control!' and I just remember us in one incredible dive in this Wellington. We lost our capacity to hold height. We seemed to just drop like a lump of lead and I recall coming out of it at 11 000 feet. We flew on to what was theoretically our target area. Navigation was hopeless so we could have been anywhere within twenty or thirty miles of Hamburg when we dropped our bombs. I think many others did likewise.

Peter Balderston has vivid recollection of a storm over Hamburg on 2/3 August:

We bombed on ETA really and we didn't see a thing. We had a lot of static electricity around the aircraft and you couldn't see out because of the flashes.

I went back to the astrodome and there was a metal clip in the top for the navigator to hang his sextant on and there were flashes of light going from the frame of the aircraft as the lightning hit it.

Looking out of the astrodome you could hardly see a thing. There were streaks of light coming from the Wellington's wingtips extending back possibly fifty yards and the two propellers were just discs of light.

Every now and then you'd go by a cloud and you'd see a great flash of light from the aircraft to the cloud or from the cloud to you. This lasted about four hours.

The met had made a bit of a blue because they said that this storm was ahead of us and it was moving and we would be behind it all the time. But we got about halfway over the North Sea and ran right into it and were in it until we were on the way home.

We didn't know whether the flashes were lightning or anti-aircraft. We were hoping it was lightning all the time. Other than the storm it was uneventful.

I tipped the window out of the flare chute and it was a bit hairy being down the back all the time.

Pretty scary when you were throwing it out. We had to do it from the back flare chute because they hadn't installed the front chute at this time. I had to go back there in the dark with all this lightning flashing around and not knowing what was going on. Had to use oxygen by plugging in to a socket.

Harris, as well as introducing the bomber stream or concentration of forces in attack, laid emphasis on the

11. Cu-nimb — abbreviation for Cumulo-nimbus. These are thunder clouds, which were avoided wherever possible because of the extremely violent air currents in them. Aircraft had been known to break up or be thrown out of control in such clouds.

Bombing up a Wellington.

use of incendiaries on the principle that it was easier to burn an area than to blow it up. This technique was used to deadly effect on the Hamburg raids.

Bob Henderson's story of his sortie shows how the effect of the fires worked to his advantage:
This was the last of the four Hamburg raids. I had been absent for the first two. I had been playing cricket at Lord's where I was selected in the team because Keith Carmody was missing on operations and I got the call. Hughie Edwards insisted that I go down.

I didn't want to go down to play cricket but he insisted.

While I was away a sprog crew was short a bomb aimer and a wireless operator for the first raid and my two boys volunteered to go. The one aircraft missing happened to be from our squadron. When I got back the news of their loss confronted me.

We got a bum steer from the met. A front was supposed to have cleared before we hit it.

It was a most devastating experience to fly through cumulo-nimbus at 25 000 feet. The noise was fantastic. Everything metal in the cockpit glowed from the St Elmo's fire. The throttles and that sort of thing, the metal beading around the window was dancing with light and in the centre was a tiny area of clear space — just left of the frame of the window. Everything else was dancing with wretched lights.

There was the noise of thunder, and in the midst of all, we were in icy cloud and one engine started to splutter and then another. We were obviously in dire trouble. We were losing height pretty rapidly from 21 000 feet to 11 000 feet but, as luck would have it, we came out over Hamburg at 11 000 feet.

The cloud base had risen clear of 11 000 feet because fires from the three nights were burning and the heat generated had raised the cloud to that point. We were very lucky and bombed exactly on target and with that

heat the two motors which were spluttering due to icing and had been pretty well useless, gradually came back into action. Until then we were ready to jump and although the boys were ready to go I hadn't actually ordered them to do so. It did look as though we had no chance of coming out of it.

Arthur Doubleday on Hamburg on 2/3 August:
A memorable raid was on Hamburg. We did it in very dirty weather and got a thrashing. It was no more difficult than Essen or Cologne but it just stayed in my mind.

I was thoroughly iced up and was belted with flak while I was trying to deal with the ice. We had to get through a searchlight belt which stretched from Bremen to the south and into northern France.

It took normally about fifteen minutes. They either picked you up with searchlights and brought the fighters up or they threw flak into the cones of the searchlights. That was the sort of belting I meant.

As a result of the introduction of window as a countermeasure, the Germans could no longer use their early-warning devices as effectively as previously. Consequently a new strategy had to be introduced. The ground control system was, to a great extent, phased out and two new tactics were introduced — wild boar ('Wilde Sau') and tame boar ('Zahme Sau').

The wild boar strategy relied mainly on visual sighting. The aircraft were normal day fighters and the pilots relied on visual sighting of bombers in the target area which was lit by flares and searchlights.

The tame boar strategy used radar-equipped fighters. They were directed by ground radar to the bomber stream which was readily found by its telltale trail of window. Once in the stream the fighters used their own radar to track down targets. It had not been appreciated by the RAF that the radar transmissions from the H2S navigation apparatus could be used by German fighters to home in on bombers. With their superior speed and armament the fighters made many kills.

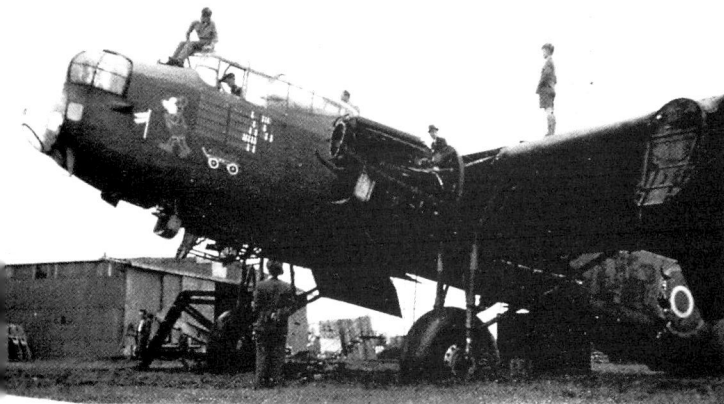
Lancaster bomber.

The Battle of Berlin — 18/19 November 1943 to 31 March 1944

Berlin, the German capital, was only fifty miles from the Polish frontier and involved a long and dangerous mission for attacking aircraft What is known as the Battle of Berlin commenced in November 1943 and lasted until 31 March 1944.

The total number of sorties during this period was just over 9 000 and the number of aircraft lost was 492 during the actual raids, while a further 95 were damaged beyond repair; a total loss of 587 aircraft. The majority of the attacking aircraft were Lancasters. They were accompanied by a small number of Halifaxes and a small number of Stirlings.

Lancaster of Queen's Flight visiting Imperial War Museum, Duxford, Cambridgeshire

Sam Weller tells of the feelings of crews when briefed for Berlin:

The problem in a 'big city' trip was the fear before you got there rather than the getting there.

When you went into the briefing room and saw the ribbon stretched across and down to Berlin you'd say, 'Ah! Look at that!' A target nearby would not cause the same reaction.

You built up this fear. It was always going to be bad going to Berlin whereas, if you were going to Leipzig, for example, it might be a bit better. I don't think it was.

Ted Eagleton tells of one of his trips to Berlin:

On 29/30 December 1943 there was an extremely large attack. I was flying a Halifax and it was a relatively easy trip although there was heavy opposition over the target as usual. The met. was spot-on. My navigator's oxygen became detached and he passed out at 24 000 feet. I cleared the target and had to immediately drop down to 12 000 feet where the engineer revived him with oxygen bottles.

He had given us the homeward course before passing out, so we simply flew that heading.

We stayed at 12 000 feet. This put us in the low flak zone and we were quite isolated from the rest of the stream.

I didn't want to climb back up to the rest as I wasn't sure I would have enough fuel.

Arthur Doubleday experienced a similar oxygen problem:

This was the first trip to Berlin I did with my crew. The first thing that happened was that at about 1000 feet the airspeed indicator fell to nil. We went on without it.

The navigator, however, ultimately got us to the target on time.

En route I asked him how long to go and he told me fifty-eight minutes. A quarter hour later I asked him to check the time and he told me eighty-eight minutes. The wireless operator said, 'We're doing bloody fine on our first trip, we're going backwards.'

But the navigator was sitting with his chair on his oxygen tube. He had cut his supply and was almost unconscious. We gave him some oxygen and he came good. He obviously hadn't been navigating since he gave me the turn on hitting Denmark.

Peter Balderston, who was in Geoff Coombes' crew (see next story), has unhappy memories of Berlin — he was shot down:

I went to Berlin in a Halifax between Christmas and New Year 1943 and it was just like a cross-country run. We saw flak, we saw some fighters, but nothing near us. We flew to Berlin, dropped our bombs on the target indicators and came home. It was a beautiful trip of about seven and a half hours.

The next time I went was on the night of 28/29 January 1944. We had to carry a second dickie — a pilot going over on a raid for the first time. His name was Doug Hughes. We flew over the North Sea and across Denmark to one of the islands. We had to turn practically south east to Berlin — a long leg to Berlin.

In those days we would aim for about three different towns to fool the Germans.

Countdown to a sortie: crews approach their aircraft, Leconfield, 1943.

On this raid the Germans were right onto us. As we were going across Denmark the port outer engine was knocked out by flak. Then a piece of flak hit the second dickie right on the parachute buckle and laid him out cold and clean as a whistle. We dropped his seat, took him out, woke him up and made sure he had oxygen. He seemed okay after that.

We pressed on to Berlin. As we were approaching the target a Ju-88 caught us and, from what I can remember, he may have made six or eight attacks. Cannon shells were going everywhere. The rear gunner was firing but after a second or two I didn't hear any more firing from our guns. The Jerry must have got the rear and mid-upper gunners. A cannon shell hit my wireless and it blew up. Later my German captors dug pieces of it out of my hair. I must have looked a horrible sight. I was hit under the chin. Another cannon shell hit the oxygen bottles right behind my seat and they blew up with a terrific bang. I thought he'd got the bombload.

By this time Geoff had called, 'bale out!' so I helped the navigator on with his chute. He got out ahead of me through the escape hatch. He was killed — whether from wounds then or later we don't know.

The bomb aimer was wounded too. Our two gunners did not survive. I spent the rest of the war in Germany.

Geoff Coombes gives his version of the attack:
I could see the German guy in the cockpit — that's how light it was. I could practically see the rivets in the plane. Standing at about 800 yards, just out of range, his wing dipped and this indicated he was coming in towards us so I took a dive towards him which gave him maximum deflection. This time I don't know if he was a bit early or I was a bit late but his guns flashed. The tracers had previously been veering across the sky because they had been missing. This time they came in straight.

I was wounded in the head and passed out for a few minutes. This night we had taken on a second dickie but we had no dual controls. The crew tried to get me out of the seat because they thought I was dead.

The attacker came in again from the other side and gave us the works. We lost two engines and half the perspex out of the front. The bomb aimer got a bullet in the back.

We had two motors — I had feathered the other two. I was a bit scared. I didn't know where the bloody German was.

Just after that the intercom went dead and none of the crew could acknowledge the call to bale out.

The mid-upper gunner came past me and they all started to pile out of the front hatch.

The rear gunner would normally go out the rear so

I had to presume he had gone.

The only communication was a little indicator light. I was halfway out of the seat when this light came on in my position. Obviously the gunner had not left so I got back in my seat.

We had a call sign for baling out — a series of Bs in Morse code and anyway I sent him a series of Bs to bale out and I couldn't get any answer back.

I kept the nose up and rode it down. We were going down in a big spiral. I stayed with it to 7 000 feet to the end of the cloud layer and I baled out. He didn't survive. He must have been critically wounded.

I was in hospital for six months.

Luckier than these crew members was Ted Priest who, on a trip to Berlin, 'had an enemy fighter make a pass at my plane. I had a go at it but it was that quick that before I had a good burst it had disappeared.'

Enemy fighters did not always press home attacks.

Other Targets

During the Battle of Berlin period there were, of course, other targets, and of these I intend to cover only four — Frankfurt-on-Main, Magdeburg, Leipzig and Nuremberg.

Frankfurt-on-Main was an industrial area specialising in machinery and chemicals. On 20/21 December 1943 a heavy raid of 576 aircraft dropped 2070 tons of bombs. Nearly all of them fell outside the city and the result was minimal. Forty aircraft were lost.

The Australian 466 Squadron sent fourteen aircraft on this raid. It was the squadron's first raid on Germany with Halifax III aircraft Two aircraft were lost. One was flown by Flight Sergeant Pat Edwards. Bruce Loane and Ralph Parsons were members of this crew and survived the attack and the subsequent period as POW's.

Bruce Loane:

We were actually shot down because Bomber Command had a new scheme on. The idea was that, after bombing, we were to turn back over the target. I don't know the reason for this move.

We were attacked by a fighter but at the time I didn't realise it because I was off the intercom. I was jamming out the ground-control broadcasts to the fighters who were in the area. If directed from the ground they could be vectored onto us and we were sitting ducks. We were one of the early Halifaxes with H2S on that night.

We were not getting much joy from the H2S and the pilot was wondering whether we should keep it on. We were all conscientious and decided to persevere with it but I don't think we were getting any result from it.

I remained off the intercom as I was to stop the

instructions coming from the ground. The Germans had a number of frequencies they could use and these were known to intelligence who gave them to us on take-off. The Germans would switch to various frequencies as we interfered with them and we had to follow and jam them out.

Being busy, I was not entirely aware when the first attack was made. I understand it was a Ju-88. I could only switch into the intercom every now and then. I knew we were in trouble but I probably missed the pilot's instruction for the back part of the crew to bale out.

I was underneath the pilot. I could see the bomb aimer was still prone and the navigator, who was a little bit ahead of me.

All of a sudden, the bomb aimer got up quickly, grabbed his parachute and clipped it on. (We had chest parachutes in those days). He almost knocked the navigator off his seat as he opened the hatch. He had a bit of trouble opening it.

At that stage I abandoned the wireless exercise because we were obviously on fire. Then I was talking to the pilot and we were getting into a deep dive and the wings were starting to drop. I kept asking whether we could go to France — which was ridiculous because we were a long way from there. I was clutching at straws. So I grabbed my chute and clipped it on. I had

Bruce Loane and crew, 466 Squadron RAAF Driffield, Yorkshire. From left: R. E. Spalding (air gunner) G. B. G. Brett (navigator), Pat Edwards (pilot), Bruce Loane (wireless air gunner), Ralph Parsons (bomb aimer), Jim Cahir (air gunner).

trouble getting out of the hatch because we were in a very steep dive. The centrifugal forces were pretty bad but I got out.

I turned over a couple of times in the air and I was worried about coming down in water because it was night and very dark.

More by good luck I was right way up when the parachute opened and I was able to use the shrouds. The aircraft crashed just near me and the pilot was killed. I crashed into the ground and was knocked out. That was that.

Ralph Parsons was the bomb aimer in the same crew as Bruce Loane but he saw things slightly differently:
Suffice to say that our mission was successful until setting course for home. We were attacked from below by a night fighter probably using upward firing cannon — referred to by the Germans as 'Schrage Musik.'

The starboard inner engine was soon burning fiercely and racing out of control. The controls were useless and the aircraft went into a steep dive and made a deafening screaming noise. The fire was spreading fast and I could see that the wing framework was red hot and pieces of metal were flying back off the wing.

Our pilot, Pat Edwards, gave the order to bale out.

After making sure that Pat's parachute was readily accessible I made my way down to the escape hatch under the nose of the aircraft The rushing air made it difficult to get out. I curled up in a ball and rolled out, turning somersaults. As I rolled over and over I was

Bruce Loane with the Luftwaffe pilot who shot him down.

unable to pull the ripcord owing to the proximity of the aircraft It eventually sped away leaving a fiery trail.

I pulled the rip-cord and the parachute opened with a loud bang. I must have been falling pretty fast by this time and the opposite reaction gave me the sensation of speeding upwards. I thought I was off to Heaven (a most unlikely possibility). Then such a silence that I could hardly imagine possible.

German crew which accounted for Bruce Loane's 466 Squadron.

Being a lightweight I seemed to be suspended in space for ages. While floating down I saw the aircraft crash in flames probably four or five miles away. Unfortunately Pat was unable to make it. Being last to bale out, he would be unable to make his way through the fire. Pat was just twenty one years old.

Eventually I could see the silhouette of trees etc. on the horizon and I braced myself for a landing. I landed on the side of a small hill. My chute got tangled in a fence just above me.

Ted Priest, a gunner on 466 Squadron, warded off an attack by a German fighter on this sortie:
The Germans were dropping flares on our path when a Do-217 [night fighter] loomed up in the darkness. The bomb aimer saw it first. We passed it and it was high in the air. I had a shot at it. The bomb aimer reckoned he saw sparks coming out. Anyway, we couldn't see it go down in flames but we claimed it as damaged.

Other crews reported they saw an aircraft firing at a Do-217. They reckoned it was damaged but I couldn't claim it as destroyed.

Magdeburg was a target connected with the oil industry. It was well into Germany.

The main stream was now quite large and in poor visibility the risk of collision was a hazard which beset pilots.

Magdeburg was attacked on 21/22 January 1944. Ted Eagleton takes up the story:
Everything was going fine. We were flying through some cloud and just as we were reaching our second level — it was one of those trips when you flew to one level and then to another — I got a smack in the tail from another Halifax.

This jammed my port rudder and in the finish it was just swinging by a top hinge and the elevator was damaged as well. The other plane had taken a cut out of the rear turret and I presumed either a wing or propeller had hit it.

Well, we tried to go on but we had to reduce engine revs in the two port engines to try and overcome the drag of the aircraft and we decide to give it away. We turned around — this took a long while but we eventually got it round — and set course for home. We were pleased to finish the long flight over the sea and it was pretty difficult to get down.

I had the rudder controls tied up with a piece of rope. I always carried a piece of rope in my pocket — always. I handed the rope to the bomb aimer and he tied the rudder bar to a strut to take the weight off my legs. I couldn't hold the rudder on my own.

After quite a long approach I got it down and we were down safely after about four bounces along the runway. We landed at Leconfield.

Recapitulating, loss of rudder meant an awful lot of drag on the side of the aircraft so you couldn't keep the aircraft straight. Having reduced power on two engines helped to control it but virtually the aircraft was crabbing along so we took quite a big arc getting back home. We dropped our bombs over enemy territory and some farmer probably got them.

Leipzig is situated on the North German Plain, on the border of Saxony. It is deep in Germany and a long haul from the airfields in England. It was one of the targets lumped into the Battle of Berlin.

Arthur Doubleday flew in an attack on 19/20 February 1944 and he recalls:
In February 1944, towards the end of which was real winter stuff, I can recall Leipzig. The take-off was delayed and the tactics changed. We were supposed to go out low, then climb suddenly and go in at Texel. My gunner reported seven explosions or flashes over the North Sea before we got to Texel. It looked like seven collisions which would have involved fourteen aircraft I had warned the pilots of my squadron to be careful about cutting across the stream but it looked as though some crews had come to grief.

When we did reach enemy territory the weather was clear. We'd been through all the rough stuff.

Exactly when I got my nose over Texel the German fighters laid two lanes of flares as far as I could see and these kept spinning around — the wild boar and tame boar aircraft were with us all the way to about Hanover. Then they either got tired or ran out of fuel. After Hanover the intensity of the fighter attacks diminished but there were streams of burning Lancasters and Halifaxes on the ground to show the way to go.

The target itself was very brightly lit with searchlights. But after we got away from that it was a quiet trip home. Seventy-eight planes were lost that night.

That was another so-called maximum effort.

Ted Priest was on this raid and saw the sortie from a rear turret:
It was a stuff-up because we arrived half an hour before the master bomber. We had a tail wind instead of the forecast head wind. Everyone arrived before the flares were dropped and had to circle around. There were collisions everywhere.

The initial met forecast for the Nuremberg raid of 30/31 March 1944, was for high cloud cover and clear weather over the target. The raid proceeded although a subsequent met report from a Mosquito aircraft cast doubts on the original forecast. The Mosquito was right.

Of 795 aircraft that took off, ninety-five were lost. It was the heaviest loss of the war. A German fighter tactic used with success on this occasion was the fitting of fixed upward-firing cannon in the fighters. The fighter could then attack unseen from underneath the bomber, which was blind in that area.

The raid was a failure: 120 aircraft bombed Schweinfurt in error; the remainder did little damage to the target.

This was the end of the Battle of Berlin.

Arthur Doubleday tells of his experience:

Nuremberg was another so-called maximum effort. In his book '*The Nuremberg Raid*', Martin Middlebrook wrote that there was tremendous apprehension. It was not the situation as far as I was concerned. It was a normal day. We turned up at eight o'clock and went into the normal routine.

It was not violent weather but rapid changes of weather. Forecast winds were about 120 mph and I think they were about half that. Lots of people got over into the Ruhr.

When we were running the aircraft up it was perfectly all right. When we started up to taxi out the distant reading compass didn't work. I said to the crew, 'With the exception of the engineer you've all done a full tour without this so you can just use the P6.'

So off we went but I didn't have my own navigator. This 'spare bod' navigator had done twenty-nine trips but I don't know to this day if he took the deviation card for the P6 or the deviation card for the distant reading compass. I think that he was off course about three or four degrees to the north.

We crossed the coast of Belgium and were then on the long leg as we called it — 280 miles I think. It was as clear as a bell, not a bit of cloud cover. The outside air temperature was -56°C. So there were condensation trails everywhere. You could just get on to a condensation trail and follow it.

A funny side. With -56 deg. outside and a nice warm atmosphere inside — about 25 deg. — the warm air misted up the windscreen which iced up. For sixty miles I spent most of the time scratching the ice, which was about a quarter-inch thick inside. My fingers were numb so my scratching became ineffective and I couldn't see where I was. I was at my wit's end.

The rear gunner was calling, 'port!' 'starboard!' the mid-upper gunner was giving some information, the wireless operator was saying, 'I've got a contact 1500 yards.' I was just about to say to the engineer, 'Wipe a glass of coffee over the whole of the windscreen and see if it will clear.'

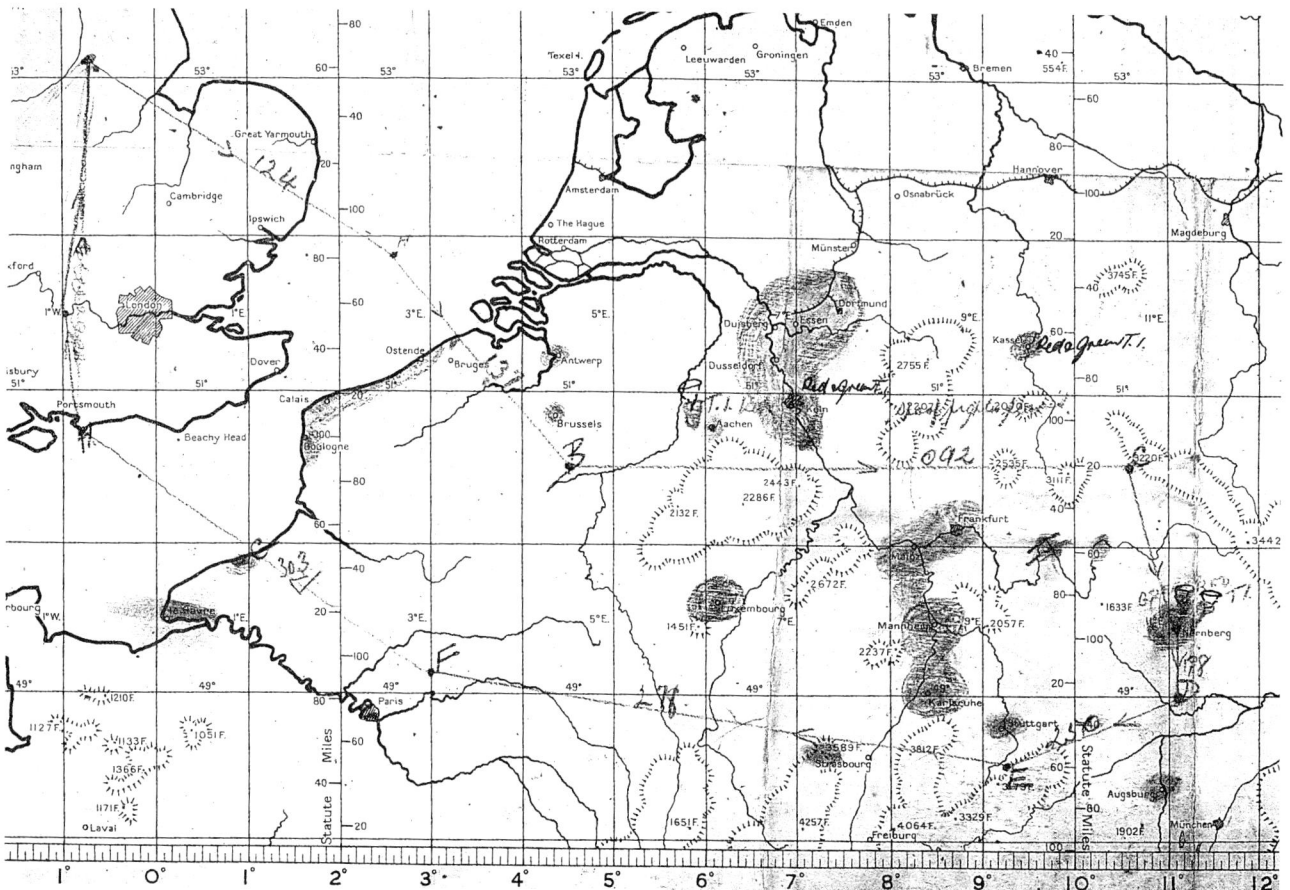

Pilots map used by Wing Commander Arthur Doubleday of Nuremberg raid, 30/31, March 1944.

I suddenly thought' 'I know where there's hot water.'

So I took out my handkerchief and urinated on it, wiped it round the windscreen and Presto! The ice disappeared. There wasn't a bit left and it never came back. So when I spoke to the met man after I got home he said, 'Of course it wouldn't. The ammonia in it would stop the ice.' I thought, 'I'll get a patent on this.'

From there on I had a clear screen.

We got raked with fire as two fighters rolled over towards me. The mid-upper gunner yelled 'Dive!' and I didn't know where the other fighter was so I almost half-rolled the Lanc and went down. When we got away from them I said, 'Look! You should have said "climb" because diving fighter aircraft don't go up.' Anyway the attacks eased off on the way home.

When we got back Sir Ralph Cochrane asked how we went. I said, 'I think Jerry got a century before lunch.' He didn't, he only got ninety-five.

There was a funny side. I was acting CO at 467 during January, February and early March and one of the duties was to interview people for commissions. A young gunner was lined up for the interview and I had a look at his record. I said, 'You're doing very well indeed. You only turned twenty a few weeks ago and you've done five trips. My advice to you is to come back in six or twelve months time and I'll be very happy to support your application.' Six months was really a long time and twelve months interminable. Anyway, when I came home from Nuremberg, I was putting my flying gear away and this young gunner came up to me and said, 'Excuse me, I would like to reapply for my commission. I'm twelve months older than I was yesterday.'

Ted Priest commented on the losses — the chop rate — suffered in 1943:

1943 was a bad period. Especially you'd note that the younger crews who came in after you would only last a couple of weeks and they'd be gone. New crews were coming in all the time.

Lots of crews were lost on 58 Squadron. I was very fortunate that when we went to Burn and finished my last three trips I just finished before the Nuremberg raid which cost so many crews including five from our squadron.

There was only one Victoria Cross awarded during the war to a Halifax squadron. It was awarded to Cyril Joe Barton RAFVR of 578 Squadron for his heroism on the Nuremberg raid.

Ted Priest (RAAF), who was on his squadron, provided me with the citation from the London Gazette of 27 June 1944.

On the night of 30 March 1944 PO Barton was captain and pilot of a Halifax aircraft detailed to attack Nuremberg. When some 70 miles from the target the four-engined bomber was attacked by a Ju-88. The first burst of fire from the enemy made the intercommunication system unserviceable. One engine was damaged by a Messerschmidt 210 which joined in the fight. The bomber's mid-upper guns were put out of action and the gunner was unable to return the fire. Both continued to attack as the bomber approached the target and in the confusion caused by the useless intercom at the height of the battle a signal from the captain was misinterpreted and the navigator, bomb aimer and wireless operator left the aircraft by parachute. PO Barton faced a situation of dire peril. His aircraft was damaged, his navigational team had gone and he could not communicate with the remainder of his crew. If he continued his mission he would be at the mercy of hostile fighters whilst silhouetted against the fires and flares in the target area. He also faced a four and a half hour journey home across heavily defended enemy territory. Determined to press home his attack at all costs, he flew on, reached the target and released the bombs himself. As PO Barton turned for home, the propeller of his damaged engine, which was vibrating badly, flew off.

The flight engineer discovered that two of the petrol tanks had suffered damage and were leaking. PO Barton held his course and, without navigational aids, and in spite of strong headwinds, successfully avoided the most dangerous defence areas on his route. Eventually, he crossed the English coast only 90 miles north of his base. By this time the petrol supply was nearly exhausted. Before a suitable landing place could be found, the port engine stopped. The aircraft was now too low to be abandoned successfully. PO Barton signalled the three remaining members of his crew to take up their crash stations. Then, with only one engine working, he made a gallant attempt to land clear of the houses over which he was flying. The aircraft crashed and PO Barton was killed but his three comrades survived. PO Barton had previously taken part in four attacks on Berlin and fourteen other operational missions. On one mission two members of his crew were wounded during a determined effort to locate the target despite appalling weather conditions. He gallantly completed his last mission faced with almost impossible odds. PO Barton displayed unsurpassed courage and devotion to duty.

"NO ENEMY PLANE WILL FLY
OVER THE REICH TERRITORY"
HERMAN GOERING

Lancaster from 467 Squadron RAAF with 98 raids painted on it and slogan quoting Goering:

MINING OPERATIONS

Problems with its Beaufort aircraft forced Coastal Command to hand over this responsibility to Bomber Command.

Harris had been a prewar supporter of the use of bombers to lay mines. His No. 5 Group with its Hampden aircraft were engaged as early as April 1940 in minelaying — an activity which soon became known as 'gardening.'

Some 50 000 mines were dropped by Bomber Command during the war and they accounted for 40 per cent of Germany's shipping losses.

John Page, Arthur Doubleday, Ted Eagleton and Peter Balderston flew on gardening duties and tell of some experiences.

John Page did his gardening in Hampdens:

I guess 'gardening trips' were interesting enough. They involved dropping mines from low level into the harbours of St. Nazaire and Lorient, for example, to try and block

Lancaster bombers dispersed in the snow at Waddington, Lincolnshire.

the entrances to the U-boat pens. We dropped them from about 150 feet to get them as close to the pens as possible.

Hampdens weren't so easy to fly at that height. You used your altimeter but had to watch the waves to check your height. Occasionally you could get a line on height by watching the fishing boats.

Arthur Doubleday:

The trouble with mines was that you couldn't drop them below 400/500 feet. They could not withstand a high-speed impact with the water and were fitted with a small parachute to slow them down. This needed a few hundred feet to be effective. On the other hand, if dropped too high, the chute was not big enough to stop the mine falling too fast.

Ted Eagleton flew a Wellington on mining trips:

We did a trip to Brest one night — a mining trip. I did two or three during my tour. On 6 August 1942 this was an extremely hot trip for mining. We had to pick up a pinpoint and make a timed run from there and drop our mines.

On this particular night we missed our pinpoint on the cape and before we knew where we were we were flying over the mainland and I said, 'What's going on?' to the navigator and he said, 'We should see the island now.' I said, 'Well the island I know hasn't got train lines on it and this particular island happens to be the mainland.'

So he said, 'Turn south.' We turned south and finished up right in the middle of Brest Harbour. Mind you we were flying at about 500 feet and the cliffs there were about 300 feet high so we were extremely low in relation to the ground.

We then had to turn west to get out. All the lights were on in the docks and people were working there.

As we were going past the docks — going through the heads as it were — the flak started and we were coned. Two sets of searchlights coned us. The rear gunner was firing at the searchlights. There was crossfire. When I looked at my altimeter it was down to zero feet. This was across water of course.

I dropped a mine — this was standard procedure if you got into trouble. Immediately I did, the German gunners lost their aiming point. They could see the mine come down. The searchlights homed on the mine rather than the aircraft and we were able to get out to the open sea again and then we picked up our pinpoint and dropped our remaining mine. The requirement for straight and level flying was risky. We had a bit of flak damage on the fuselage.

Arthur Doubleday, DSO, DFC, 460, 467 Squadrons RAAF, 61 Squadron RAF.

Peter Balderston:

While on the Wellington we used to do a lot of minelaying. St Nazaire, La Rochelle, Brest and these types of targets. This would be in June/July 1943. I did about eleven or twelve trips on the Wellington and I'd say at least six or seven were minelaying.

A minelaying trip was the most hairy because you were coming in low. To get your altitude you used your altimeter which was worked by barometric pressure which varies from place to place. To get your altimeter setting you had what was called an altimeter flash which the navigator threw out of the flare chute and he timed when he threw it out and the rear gunner timed when he saw the flash. The navigator then calculated how high we were. It was not very accurate. As we were flying at anything from 50 feet to 200 feet we could make an error of 150 feet quite easily. You wouldn't want to do a steep bank because you might put your wing tip in.

We had to come in low initially and then, to lay the mine, had to go up to 700 feet to give the parachute time to open.

Command was very strict on the pinpoint because they had to know the position of every mine that was dropped. It would plot it on a larger chart.

On one of these trips we had to go in between an

island and the mainland of France — only about half a mile wide with guns on either side and we had to fly across because we had to pinpoint a spit of sand to get our position. It was a timed run from this spit and up to 700 feet to drop your mine. La Rochelle I think.

As we were going in light flak was coming from both sides of us and then, right ahead, about 200 yards away, there was a mighty explosion. One of our squadron blew up. The flak must have hit the mine. Metal from the plane hit us too.

You could see the tracers. You would swear that every one was going to hit you between the eyes. Then it would rip up over your head. Fortunately they didn't hit us.

Lancaster bombing up, 467 Squadron RAAF.

5

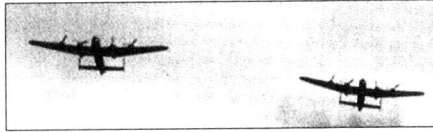

BOMBER COMMAND

1 April 1944 To 9 May 1945

BOMBING OPERATIONS

1. Prelude to the Invasion of Normandy and Associated Operations

This chapter covers Phase 3 of the operations of Bomber Command under the direction of 'Bomber' Harris.

The approach of D-Day and the re-entry of Allied Forces into France led to changes in both command and direction of both the British and the US Air Forces. In April 1944 both Air Forces were formally placed under the command of the Supreme Allied Commander — General Dwight Eisenhower. There was an immediate change in emphasis. Attacks on rail centres and prime military targets, such as gun batteries, airfields etc. were stepped up.

Alby Silverstone, a bomb aimer on 466 Squadron RAAF,

OPERATIC IS RECORD BOOK.

of (Unit or Formation) No.466 Squadron R.A.A.F.

SECRET.

Place	Date	Time	Summary of Events.	
DRIFFIELD	1.2.45	1.2.45	Squadron detailed to provide eighteen aircraft for night attack on MAINZ.	A.175
			All took off,bombed,and returned safely.Bombing was done on skymarkers	B.258
			with unobserved results,but the attack appeared to be fairly concentrated.	918-924
			Flack was only slight to moderate,and two fighters were seen.	
	2.2.45		Sixteen aircraft detailed for night attack on Wanne Eickel.All took off,	A.175
			bombeb and returned safely.The target was covered in cloud,obsuring the	B.240
			markers and bombing was done mainly on Gee fixes.Flack opposition was	
			moderate and there was one combat.	924-929
	3.2.45		Squadron not required for operations.Training flights carried out.	
	4.2.45		Eighteen aircraft detailed for night operation on Gelsenkirchen (Nordstern)	A.177
			Two non-starters and one early return-all justified. Others bombed and	B.241
			returned safely. Owing to 10/10 cloud bombing was done on Gee,no markers	
			being visible.Flack did not worry crews at all,and there were no combats.	924-934
	5.2.45		Fifteen aircraft detailed for night operation but this was cancelled at	
			1900 hrs on account of weather. Limited flying training carried out.	
	6.2.45		Squadron not required for operations.Several bombing details attempted	
			during morning but the weather was unserviceable.	
	7.2.45		Eighteen aircraft detailed for night tactical operation on Goch,several	A.178
			miles inside the German lines.All took off and returned safely. The attack	
			for the first seven minutes was concentrated,but the master bomber then	B.242
			cancelled the rest,owing to the obliteration of the markers. A fighter	
			thought to be an ME410,was shot down by F/O.Russel and F/S.Cutts,in	935-940
			F/O.McCallum's crew about 15 minutes after leaving the target.	
	8.2.45		Sixteen aircraft detailed for night attack on Wanne Eikel,which attack,after	A179
			several postponements,finally came off at 0615 hrs next morning.All took,	B.243
			but one aircraft (F/O.Gilbert) returned early on 3 engines. The bombing	941-946
			seemed to be fairly concentrated on the markers,and several large explosions	
			were seen. The defences put up a moderate barrage.	
	9.2.45		Squadron required to provide six aircraft for night attack,which was	
			cancelled. Several	

Sheet from Operations Record Book, 466 Squadron RAAF, Driffield, Yorkshire, 1-9 February 1945.

flew his first trip at this time — to attack a gun battery on the French coast:

Although we knew that we could not have an easier trip on which to open our tour, there was a lot of nervousness in the crew. This trip was for real and, for the first time, we were carrying high explosives — 11 000 lb (5 tons) of it. Although the bombs aren't armed at the time of take-off, it was a fairly lethal load to be carrying. The tension was quite high as we taxied 'X' X-Ray out of our dispersal point to join the stream of bombers making their way to the end of the runway.

At this time only the pilot, flight engineer, rear gunner and myself were at our normal stations. The navigator, wireless operator and mid-upper gunner were seated on the floor in the centre of the aircraft, with their backs braced against a bulkhead. As a bomb aimer, I was occupying the second pilot's seat, but after take-off I would enter the nose of the aircraft to help with the navigation and prepare for the bombing run.

The decision when to leave the dispersal point and join the taxi path to the runway was left to each crew as radio silence was imposed to conceal from the enemy, as far as possible, the fact of the take-off and the size of the force involved. Take-offs were controlled by Verey flares and signalling lamps. There was no radio silence for the return to base.

Slowly we made our way to the take-off point, turned on to the runway and, when the green Verey flare was fired, Aub opened the throttles and released the brakes and we raced down the runway. He called out, 'Okay. Hold them tight!' and as I held the throttles wide open, he concentrated on getting the aircraft airborne.

After we had circled to achieve our required height we set course in the enveloping darkness, down the length of England with seven pairs of eyes swivelling in all directions. Shortly afterwards, I climbed out of the second pilot's seat, took my parachute pack from the wall bracket and made my way to the nose to take up my bomb-aiming duties.

My first action was to test the bomb-release mechanism and then set the order of release for the various bombs, each of which was in a different station with its own individual release mechanism. I then moved back behind the black curtain and took up my position at the navigation table.

From time to time I set up a fix for Jim using the Gee set and, as we progressed towards the target, I checked the course from the H2S — a new form of radar navigation aid which recorded the topography of the ground on a screen.

This system, which had been only recently introduced, transmitted radar beams downwards and received reflections back from the ground. No reflection was received from water, a small amount from flat ground but significant amounts from buildings, hills, forests and mountains. Thus a picture could be made of the ground below the aircraft. The centre of the picture was the aircraft and the edge could be set at ten or fifty miles. An experienced operator could read his way over most terrain that contained cities, towns, rivers or coastlines.

When about ten minutes from the target I moved through the curtain, parachute pack in hand, and settled over the bombsight waiting for the target to come up.

As we approached we could see searchlight beams waving about as they looked for the bombers. Then the coloured target indicators fell from the sky to mark the target. We were still some miles from the target but we could hear, over the VHF radio, the master bomber giving the directions as to where we should bomb.

As I started lining the target up with the bombsight graticule I could see the anti-aircraft fire bursting in the air but I had little time for that as I called the course corrections to the pilot to come over the aiming point. 'Bomb doors open!' and I felt the shudder as Aub opened them. I now called the corrections to bring the aiming point sliding along the bombsight graticule and then, 'Steady. Steady. Hold that!' Then, as the target indicators reached the crosshair, I pressed the release button and as the bombs fell away in the order I had previously set, I called, 'Bombs gone!' When all the bomb station lights showed 'Clear' I called, 'All gone!' and Aub closed the bomb doors.

We continued our run for the designated time and turned and headed for home as the flak bursts flickered all round us. While we did not appreciate it at the time, it was actually quiet in comparison with what came in later raids.

On 25 July 1944, Arthur Dutch, a wireless operator on 467 Squadron RAAF, went on an operation which he entered in his diary as follows:

I really enjoyed this trip — stuck my head out of the astrodome — Lancs all round us, Gee went u/s — alternator again. Saw everything (including the flak) all the way to Villacoublay (near Versailles, near Paris). Target, an airfield and signals depot was really plastered — not one bomb outside the area. Ernie hung right above us — opened his bomb doors then dropped his load between our wing and tail. Kite flying back with us had no rear turret and a gash torn in the starboard wing. Saw quite a number of escorting Spits — no enemy fighters at all. Visibility very poor. At about 1-2000 feet when we crossed the English coast. All 467 kites had returned. One of 463 missing — one kite landed with full bomb load and no hydraulics — pranged on landing — all crew got out. Navigator had a shrapnel wound in leg.

The Ruhr

As 1943 reached its close there were some two million Germans employed on anti-aircraft duties (90 000 were anti-aircraft gunners). So, as the bomber phase entered 1944 and progressed to 1945, the anti-aircraft and searchlight defences were formidable. A good proportion of these defences were centred on the Ruhr cities.

Fred Wright, in describing his visits to the Ruhr ('Happy Valley') gives his impression of these defences:
There was a hell of a lot of flak in the Happy Valley and you just had to fly through it. But on any bombing raid you generally found that the only time you encountered the flak was over the target area. We were seldom too worried by flak except at the target. You were only there ten or fifteen minutes and then you were out and on your way home. Nevertheless the flak was disturbing, particularly at night, because you saw the flash and not just a puff of black smoke. When you went through the target area and saw these things bursting you thought, 'Crikey! How am I going to get out of that lot?' It was different with fighters. You could be attacked anywhere.

Harry Brabin, of 102 Squadron, gives a more detailed picture of time over a target — namely Sterkrade on 16 June 1944:
At the briefing, the wing commander gave us the details of our destination and said, 'Please come back. The expense of your aircraft and the cost of your training adds up to more than you can earn in your lifetime.' Pointing to a red area on the map, indicating a heavy concentration of anti-aircraft guns, he said, 'Don't be frightened of all this. Remember you will be dropping bombs on them.' It was light-hearted advice aimed at dissolving that sinking feeling. It was only too obvious that we needed to conquer our fears to be an effective crew and to return safely.

The target was a synthetic oil plant and it turned out to be a spectacular sight. Great walls of fire shot high into the air.

The Pathfinder Force was made up of experienced crews who flew over the target and dropped target indicators to identify the aiming point for the rest of the bombers. This was particularly helpful on cloudy nights.

Fred Wright, DFC, and crew of 78 Squadron RAF, Breighton, Yorkshire. *Front row*, from left: Jack Todd (air gunner), Cliff Pemberton (navigator) Fred Wright, Peter Newton (wireless operator).
Second row: Bill Donnelly (fight engineer), Terry Tregilgas (bomb aimer), Charles Tyson (air gunner).

But tonight was so clear it was hardly necessary. I didn't envy the master bomber, or his deputy, who had to fly low over the target and keep circling it while we dropped bombs. Many bombs were wasted prior to this marking procedure and landed in open fields, far from the target.

It was a great deal more dangerous to fly right over the target and bomb carefully but Don always took great care to bomb effectively; and, after dropping the bombs, we had to keep flying straight and level for what seemed an eternity so that the photographs would be accurate.

The target was visible for many miles as the searchlights came up, trying to get three on a single plane to cone it. Once coned, the chances of escaping were minimal as the enemy could work out where the plane would be and aim the shells at that spot.

Enemy fighters were everywhere. We lost our flight commander on this trip. The signals leader and the gunnery leader were in his crew. Two other squadrons in 4 Group lost six and seven planes each. Our squadron lost five.

We escaped two fighter attacks because Sandy was able to direct the evasive action and engaged the fighters with his four Brownings. A total of thirty two aircraft (10 per cent of the total force) was lost on this trip.

GENERAL OPERATIONAL ASPECTS

Flak

Jim Tudberry:

I suffered a lot of flak damage right through the tour. It was something to worry about. We didn't like it.

It sounded like hailstones on a tin roof when it hit. Some of the worst was over the Ruhr — very heavy flak.'

Sam Weller had a narrow escape:

We got hit one day. A piece of flak went right through the aircraft and went behind my seat. It made a tremendous noise when it hit and I was convinced it had made a large hole. When we got on the ground I searched for ages and ages but I never found the cause of the hole.

I always reckoned I smelled cordite when the flak was too close. When a shell burst in front of you it looked like a barrage balloon — a large black cloud. When that happened I made a sharp turn to the right. They seemed to me to fire in threes. One puff, two puffs and you had to get out of the way of the third.

Max McVicar also had a close shave:

We had done a daylight on Cologne and were heading back when, suddenly, we saw these bursts. They were so close we could hear them. A hole appeared in the perspex on the side of the aircraft.

Next day, the ground staff said, 'Come and have a look at this!' There was a hole gouged in the back of my seat.

Noel Gilmour:

We weren't one of those crews that came back badly damaged. We got a few holes, but not too many.

I had the F24 camera right under me. It had a heater on which I used to warm my feet.

One night we did get hit by quite a few bits of shrapnel and I thought, 'That was awfully close to me.' Later I found a great hunk of shrapnel embedded in the camera on which I usually placed my feet. Fortunately I had taken them off that time otherwise it would have been a case of no foot as well as no camera.

Searchlights

On the way to and from the target searchlights could be a hazard — particularly if a bomber strayed over a heavily defended area. Of course, the navigator should have made sure it didn't go there.

Searchlights probed the sky over the target to pick up the bombers. Targets were usually heavily supported by searchlight batteries. From the autumn of 1940 all major target towns had their defences strengthened by more searchlight and flak

batteries. By mid-1943, for example, Hamburg had twenty-two searchlight batteries around the city.

The blue 'master' beam was the most dangerous — once it locked on to an aircraft other beams were used to 'cone' the aircraft.

Once coned, the aircraft was the immediate target for flak. The anti-aircraft batteries could easily determine its height and course, predict its future position and fire a shell to be where the aircraft was expected to be. A shell took about thirty seconds to reach an average height for bombing aircraft, so pilots had to act quickly. Added to this, a coned aircraft was a sitting target for a night fighter.

While instant action was required of a pilot, views differed as to what was the best action. Sam Weller reacted as follows:

You knew the searchlights themselves couldn't hurt you. One of the things you fought so hard for was height, so if a beam caught you, it seemed a waste of effort to put the nose down and lose 5000 feet. The main thing was to try to deceive the enemy by slightly turning but keeping your height.

My theory was, if you went down to 16 000 feet you were going to get into the range of medium and light flak. Above that height was the heavy flak, but there wasn't as much of that.

Bob Henderson had another reaction, and a quite different evasive action:

You were confronted with what seemed to be an impossible situation. You had to apply all your training and clear thinking — if you had any left — and try to settle down and realise you were in real trouble.

On one trip I remember, I was coned. There was a great amount of light around. There were some members of the crew, for example, the wireless operator and navigator, who were virtually sheltered from the light but you, as pilot, were out there in the full glare. It was impossible to get an understanding of the attitude of the aircraft because there was no horizon; just darkness and light.

What I did was to lie with my seat fully down — right down to the floor. This took me to a point where there wasn't any direct light coming into my eyes. The fuselage of the aircraft protected me from direct light and I was then able to peer at the instruments through the wheel of the aircraft to keep it in the proper flying attitude. I concentrated on the artificial horizon and the airspeed.

Once you had determined your flying attitude you corkscrewed. You put your nose down and got to buggery out of it.

The main danger was that flak would be fired up into the cone where you were highlighted. Usually, over a target, I was most concerned about the fighters which might be there. But, if coned, I didn't seem to be worried about fighters. In fact, I think the fighters were just as much put off, or embarrassed by searchlights as we were. So I steered clear of anybody who was coned but stayed on the outside of the cone. If I saw a cone near us I would steer over near it as there seemed to be more safety from fighters there. It was not that you wanted to attract the cone to you.

This you learnt after a while. For the first ten to fifteen trips you were a sprog. However, all the experience in the world could not help you unless luck was running with you too.

Light Night Striking Force

At the close of 1943 Bomber Command was looking to provide more and more protection for the bomber force. The Light Night Striking Force was developed in late 1943.

John Page describes its function and operation during the early part of 1944:

In late 1943, Air Vice Marshall Bennett (an Australian) steadily built up this force. It was comprised of Mosquitoes and its operations were mainly diversionary and nuisance raids.

A new squadron (692) was forming and I went along to join it and complete my Mosquito tour.

The Mosquito was a delightful aircraft to fly. It was light and got up high. At the start we were carrying a 2000-lb bombload — usually a couple of 500-pounders and the rest incendiaries. Later, the aircraft was modified to carry a 4000-lb cookie, which we often took to the Ruhr.

We carried some window, but not much as we were operating at about 28 000 feet It was pushed out by the navigator, who had to also be the bomb aimer. There were only the two of us in the crew.

If the main force went to Munich, we'd go to somewhere like Cologne. We also did nuisance raids on Berlin. At this stage there were only four squadrons, so two (about sixteen aircraft) would go to one target and two to another. Occasionally they'd send two squadrons to Berlin.

I went on some purely diversionary raids. On the last night the heavies went to Berlin — in March 1944 — the squadron was briefed by AVM Bennett to fly in from the west while the main force came in from the north.

We were to fly at about 28 000 feet until about fifteen minutes from the target, when we were to drop to 22 000 feet. At five minutes from the target we were to turn on all our landing lights.

The aim was to attract any night fighters in the area to us. Bennett's instructions were to keep the lights on for a minute — no longer — and then put the nose down and go. The sole aim of the raid was to divert attackers from the main force. We lost one aircraft out of the twelve we sent.

Just to make complete nuisances of ourselves, a couple of Mosquitoes were sent to Berlin carrying a lot of window which we scattered about 100 miles from the city. The aim was to give the impression that there was a number of aircraft heading towards Berlin. We then flew over at 28 000 feet, dropped a couple of bombs and got away. We hoped that this would get the searchlights and ack ack into action, set the air-raid sirens going and force the civilians into the shelters so as to generally disrupt and demoralise the population.

On D-Day minus one we went to Osnabruck, a raid intended to cause confusion. On 7 June we went to Cologne and then Berlin as part of our normal function of raiding Germany rather than giving direct support.

On 10 April 1944 I took a 4 000-lb bomb to Berlin — the first time a bomb of this size had been taken there in a Mosquito. After that it became a habit.

I was only close to getting punished once — on the way back from Berlin one night. I was at 32 000 feet and there was a little thin cloud around. I was just stooging along when, suddenly, there was a terrific explosion under the aircraft. An AA battery must have been tracking us for some time and it got pretty close. We immediately dived down. The navigator, who had been half-asleep, had to push himself down from the cockpit roof.

We found a couple of shrapnel holes when we got back. It had been close and due to my own stupidity.

Mosquito aircraft.

We had oxygen but no pressurisation. We went up fairly fast — 4000 feet per minute was our rate of climb — and people sometimes got the bends.

One interesting raid was a low-level mining raid on the Kiel canal three weeks before D-Day. We dropped our mines from fifty feet — an unusually low level for us — and got nine into the canal. We lost one plane out of the twelve sent. It was moonlight and visibility was good. I didn't see much flak but there was a lot of machine gun fire and balloons everywhere. The canal was blocked for seven days, which was important as a lot of cargo went along it. Photo reconnaissance seven days later showed there were still more than 100 ships held up.

Some of our people were decorated as a result. The citation read:

'One night in May 1944, several crews of Mosquito aircraft were detailed for a difficult and dangerous minelaying mission. The operation called for the highest standard of skill and accuracy. In the face of intense anti-aircraft fire, balloon defences and considerable searchlight activity the attack was pressed home with great precision from low level. That complete success was achieved in spite of such hazards is a high tribute to the calm courage of the following officers who participated as members and leaders.'

Actually, our attrition rate was not too bad. But our wing commander, a New Zealander, was killed about three nights after the Kiel canal attack.

I finished this tour of fifty-one trips on 6 July 1944.

FEINTS

Another measure which was aimed to help dissipate the German fighter force was to split the main force and to use feints. This was instituted in late 1943 and became common practice for the remainder of the bomber offensive.

Harry Brabin of 102 Squadron had experience of one feint on a trip to Stuttgart on 24 July 1944:

The target was a ball-bearing factory. This was our longest trip. At the briefing, the wing commander said, 'We will play a trick on the enemy tonight. We will head towards Mannheim so that their fighters will congregate there. Then, at the last moment, we will turn to Stuttgart and bomb.' This was the fifth successive night on which it was bombed and the bombing force numbered 614.

The weather was bad. Masses of German fighters attacked, both going in and coming out. Several fighters were shot down, including an Me-109 and an Me-210. We only partly destroyed the factory but another raid, a month later, completed the job.

We lost one plane from our squadron — F/O Page and his crew — and Z for Zebra returned very badly damaged.

Feints or not, an attack on Wanne Eickel near Essen brought this reaction:

The target was in the Ruhr area where the Germans had their greatest concentration of anti-aircraft installations. The trip was a nightmare! I kept telling myself, 'This isn't happening. It's not real. It's a horror movie I'm watching. I will get out of my seat and go home in a minute.'

Harry Brabin and second tour crew. The crew members, from 102 Squadron RAF, Pocklington, Yorkshire, are: Harry Brabin (front left); John Allen, DFC (front right); Bill Rabitt, DFC, (back left); Bas Spiller, DFC (back right).

The Fighter Menace

In interviewing Arthur Doubleday, I asked an oft-repeated question: 'Were you attacked by fighters? What was it like?'

He explained that there were three types of situation: (1) where evasive action deterred a fighter from pressing home his attack; (2) where the attack was launched unsuccessfully; and (3) where the attack was successful.

In July 1944, in an attack on Stuttgart, Sam Weller experienced what may be regarded as a variation of (2) above: Coming back from Stuttgart I was attacked by two fighters. I remember the tracer going over and under the wings. When you haven't seen it before it's hard to work out what's going on. I had no idea what it was. I had only seen flak, which came up from below. Seeing tracer come over the top of the wing really shook me for a moment. I said to the gunners, 'Have you got someone firing at you from the rear?' And they said, 'Yes! Two fighters.'

Then I saw them. They passed us across the top. I was very upset as the rear gunner hadn't seen them and warned me.

They went in front of me and turned round. I put her nose down and went as fast as I could in the opposite direction.

Bob Henderson comments:
One of the things you did was to ensure that the gunners were keeping a lookout. From time to time you gave them an opportunity to look under the aircraft, which was a very vulnerable spot. Experience helped.

If a fighter felt you had seen him he would often veer off and leave you alone. He knew he ran the risk of being shot down by your gunners. If he felt you had seen him and saw you corkscrewing or taking evasive action, he would usually look elsewhere.

THE FLYING BOMB CAMPAIGN

On 13 June 1944, one week after D-Day, the flying bomb assault was launched. Four V1 rockets (flying bombs or doodlebugs) crossed from Normandy. The next attack came two days later. It was more intense — more than 200 missiles were involved over a twenty-four hour period.

Bomber Command had launched an attack on the German rocket experimental station at Peenemunde 17/18 August 1943 and then proceeded to attack the V1 launching sites in Normandy. These attacks helped disrupt the German V-weapon attack.

Harry Brabin, a wireless air-gunner with 102 (RAF) Squadron, has this to say about a daylight sortie to Foret de Nieppe on 28 July 1944. This was a raid aimed, in part, against the V1 attacks:

The wing commander began his briefing, 'A piece of cake, gentlemen. It's just over the French coast. The aim is to help our brave soldiers by bombing tank and troop concentrations and supply centres for V-rocket launches. Today we are full of surprises. Two Mosquitoes will lead this raid of over 1000 aircraft. They are equipped with Oboe and will fly on a great circle arc controlled by radio beacons from England. When they are over the target the signal will stop and they will drop flares. You will bomb the flares. Now! that's very simple. Isn't it?'

The wing commander told us that, as his crew was not as experienced as ours, we would have the honour of leading the squadron and our squadron would lead the group.

We took off and tacked on to the rear of the Mosquitoes with the command flying in formation behind us. Flying perfectly straight, we made a wonderful target for the German anti-aircraft fire at Ostend.

We were hit repeatedly and many of the holes were big enough to jump through. In fact, Don, our bomb aimer, was thinking of doing just that, and had picked up his parachute and was strapping it on when Bas called to him, 'You've got it on upside down'.

Just then, Bas was hit in the leg by a bit of flak. I tore open his trousers and held my thumbs near the wound to try to stop the bleeding. The wound was too close to the groin to use a tourniquet. Sandy came down from his turret in the rear with a large sheath knife in his hand. Bill wondered what he was going to do with it. He cut the trouser leg off — a quantity of blood had collected there — while I kept up the pressure on the vein for the remainder of the trip.

We were near the target so we hurriedly dropped our bombs, as did two aircraft behind us who had been told to bomb when we did. Because of the injury to Bas and the badly damaged aircraft, we wagged our wings and headed home. The wing commander, who was flying behind us, took up our position behind the Mosquitoes and led the rest of the bombers to the target.

We made several attempts to get out of France, but each time we approached the coast we were driven back by a wall of flak. We eventually found a relatively quiet spot, put our nose down to increase our speed and managed to cross with only a few more holes.

We saw an American Lightning flying near us so we fired off a distress flare and he escorted us back to an aerodrome — Great Ashfield — which was a Fortress squadron. Bas was taken to hospital at Ely.

On inspecting N for Nan we found there were eighty holes in it. However, it was repaired, much to our satisfaction, as we were very attached to it.

GENERAL TARGETS, APRIL TO AUGUST 1944

Doug Glasson started his second tour on 15th April 1944 and had this to say by way of comparison with his first tour:
First and foremost, the difference between the first and second tours was that the RAF were becoming masters of the air over Germany. In August 1944 there was great confidence amongst the RAF boys about the way the war was going and the way the Air Force was going. It was unlike our first tour when nobody was really confident of the outcome.

We were now delivering more and more daylight attacks and combining more with ground forces. We had Gee and H2S and, although these were not overly useful for some of our operations, the H2S still gave us significant navigational help when we went deep into Germany at night.

Arthur Dutch recorded in his diary a sortie on 12/13 August:
Our second German target — Russelsheim Opel factory. The factory makes the peoples car, Tiger tanks, Ju-88s and doodlebugs. Most of the way at 20 000 feet. Hun jammed my Monica[1] but I still got results. Bags of fighter flares all the way in and out — the track was marked for the fighters by ground flares. Plenty of searchlights and flak over the target. We bombed the first set of markers — they were three miles out. Rod Melloship missing when we came back. Ernie Gordon bombed a 'spoof'. Kenny and Johnny saw a fighter each — a 109 and a 110 respectively.

Fred Wright, a Halifax pilot, explains the risk of collisions on raids:
On three occasions I lost engines on the way to the target. On the first raid, which was on Wemars Capelle, a constant speed unit failed but, luckily, the propeller feathered instead of racing. I had to decide whether we should go on or turn back. My decision was to go on. One consideration was that we were flying among a lot of other aircraft and if we turned or dived there was a real risk of a collision. But, if we stayed where we were, we would be a bit slow but still in the stream and going in the same direction.

The principle of the stream was to put as many planes over the target as possible in the shortest time possible with the aim of saturating the defences. This led to aircraft flying at night, without lights, in close company with significant risk of collisions. So long as they all flew in the same direction at the same speed the risk was minimised. From time to time, for various reasons, many aircraft had to cut across the stream and this increased the hazard. Major areas of risk were turning points. Every aircraft was responsible for its own navigation and each would change course when it reckoned it was at a turning point. The result was a mass of aircraft flying different courses in a confined area of the sky. The turning points were a necessary part of the use of feints.

1. Monica — A radar device carried by bomber aircraft to detect fighter aircraft coming up from behind. It could give some indication as to whether the approach was from the port or starboard quarter.

Harry Redwood (pilot, centre) and Arthur Dutch (wireless air gunner, left).

Harry Redwood, Arthur Dutch's pilot of 467 Squadron, RAAF, Waddington, Lincolnshire, and the bomb he was to deliver.

An Experiment — Bomb without Pathfinders

A raid on Brunswick was carried out on 12/13 August 1944. As an experiment Pathfinders were not used. Individual aircraft picked their own aiming points using their H2S sets. Some were up to twenty miles off target. The raid was not successful. Sam Weller comments:

At briefing they told us that there was a front sitting right on top of the city. We were to bomb using H2S and, after twenty minutes, dive to about 5000 feet.

There wasn't a cloud in the sky when we got there. The night fighters were having a ball.

We found the target all right, but diving after bombing proved to be a bit of a hazard. We must have been off course. What I didn't know was that we were diving over Hamburg. We came down to the height given at briefing — about 5000 feet — and flak was bursting above us. I had an engineer who had not done any night ops but had done several daylights. He had no idea what night ops were like. In the middle of the Hamburg flak he asked me to hold the aircraft steady so he could complete his log.

LONG HAULS

Some targets, such as the Ruhr, were disliked because of their heavy defences. Others were unpopular because they were too distant and involved long hauls over enemy territory.

Always present was the likelihood of fighter attack and much determination was necessary to counter the effects of up to eleven and a half hours in the air.

Flying Officer Alan Stutter tells of one such occasion when he went to Konigsberg on 29/30 August 1944:

Konigsberg was in the USSR. On 29 August 1944 my crew and I knew nothing of it, except that it was at the eastern end of the Baltic Sea and had a bad reputation. The little that we did know came from the crews of our squadron who had raided it a few weeks earlier. They had received a hot reception and were not impressed by the distance, which had involved flying times of eleven hours or more. To add to their discomfort, the raid had not been very successful.

Given this, it was not surprising that there was some lack of enthusiasm when Konigsberg was announced as the target for the night.

To avoid the main German defences we would fly down the middle of the Baltic on both inward and outward journeys. We would carry a full petrol load and one

Lancaster bomber.

2000-lb bomb, with the aircraft at maximum weight for take-off.

It was still light when we left and flew, at less than 1000 feet, north east across the North Sea towards the northern tip of Denmark. We kept low to stay under the enemy radar and, as all the aircraft were nicely bunched up, must have made an impressive sight — not that we saw any ships.

Fifty miles from the Danish coast we started a maximum rate climb to get to 15 000 feet for the trip across Denmark. This was uneventful. It was now quite dark and we settled down to cross the southern tip of Sweden. The Swedes were, of course, neutral and their reaction could not be guaranteed. I must say they earned some friends that night.

The Swedish flak started as soon as we crossed their coast. However, it was most noticeable that it was carefully aimed well away from us. In fact they had arranged pairs of Bofors guns along our track. These were aimed outwards from us and periodically fired bursts of tracer. Navigation simply consisted of flying along the centre of the Vs of tracer. Someone had obviously done his homework. There was a great sense of loneliness when we passed from Sweden to the blackness and hostility of the Baltic.

After a long grind we turned in towards the target. Unfortunately, the Pathfinders had not located and marked the aiming point so we had to circle while they got their act together. This was not a healthy occupation since the Germans were displaying a certain shortness of temper. They were operating a large number of searchlights and considerable flak.

Our aircraft had just completed one circuit when we were suddenly caught by a searchlight. It was a generally accepted rule of thumb that it took fifteen seconds or so for a shell to get to our height so the golden rule was to alter course and or height, significantly at rather short intervals.

This I did, with some urgency and vigour, hoping there were no night fighters about — though, given the amount of flak, it looked unlikely that there were. It was important that there was no predictable pattern to my manoeuvres so I reckoned that if I didn't know what I was going to do next it was unlikely the Germans would. It seemed to work, as the light lost us after a few minutes.

We did another couple of circuits and had another brief encounter with a light before the controller called us in to bomb. No time was lost in complying with that instruction and it was with considerable relief that we headed for the darkness away from the target. Settled on a course for home, down the featureless Baltic, we kept a wary eye — or ten eyes (the navigator and wireless operator were ensconced in their 'office') to be precise — open for fighters.

After about a half hour, rippling blue lights began flickering over the perspex of the cockpit. It took me a while to realise that it was St Elmo's fire. We must have been flying through electrically charged clouds. Then the aircraft began to move about as it struck turbulent air. Flying now required some concentration, but at least the electrical discharge ceased.

However, it was replaced by the loud rattles of solid objects hitting the fuselage not far behind my seat. Ice was forming on the propellers, breaking off and hitting the fuselage. If it was forming on the propellers it was forming on the wings. The big question was, how much was forming and how quickly? Without any information on the cloud distribution or temperatures one had to guess at what was the best course of action. I finally decided to try to climb out of the icing layer.

I reckoned that, even if I was unsuccessful, the rate of build-up should decrease with height, and the higher we were the more time would be available to try alternatives if the worst came to the worst. In the event, the rattling stopped and the air became calmer at 20 000 feet. The aircraft was soft and a bit waffly at that height, it was cold and fuel consumption was on the high side, so after an hour or so I decided to come down again. Back at 15 000 feet things weren't so bad so we continued on our way.

Dawn broke as we were off the northern tip of Denmark. It revealed a solid sheet of cloud below us and another Lancaster a few miles off our port bow — also heading home.

I was now starting to feel rather jaded but there was nothing else for it but to keep going. We finally touched down after ten hours and twenty minutes in the air; for the whole of which I had sat, unrelieved, in my seat with the parachute and straps steadily getting more and more uncomfortable. Of course, the two gunners were in the same situation and were without the mental stimulus that I had had.

I have a notice in my log book:
> From A.O.C. No. 5 Group to all 5 Group bases and stations.
> Following message received from A.O.C. In C. Begins. Congratulations to all concerned in the devastating attack on Konigsberg. In spite of the severe weather difficulties encountered the entire city has been virtually destroyed at that immense range by a comparatively small force. Ends.

ARMY SUPPORT

John Thomas of 102 Squadron (Halifaxes) describes an attack to support the Army in the field.
Evening, 7 August 1944, Target — Battle Area Position Three.

The British and Canadian Army Groups were having trouble breaking the German defence of the Falaise Gap, somewhat east of Caen. A previous raid had failed to make any impression.

The first wave was 4 Group and 102 Squadron was at the front of the first wave of some seventy aircraft. We approached the target in a generally easterly direction and dusk was just closing in as the markers went down. We were on course at 12 000 feet and the line of sodium lights marking the Allied front line slid by underneath us. We were in the first twenty or thirty aircraft to bomb, and, as we held course for the photo, the noise of the explosions could be clearly heard, the aircraft controls became very mushy and we began to lose height.

My first reaction was that we had been hit by a series of flak bursts and were on the way down. Then the bomb aimer got excited about the huge explosions on the ground. Shortly after, the rear gunner excitedly confirmed the bomb aimer's claim.

As we cleared the target and began a turn to port, I heard the master bomber cancel the rest of the attack. I could see a sea of explosions covering many acres. A huge pall of smoke rose from the area and the explosions just kept creeping forward over a huge area.

Some weeks later, on leave in London, the bomb aimer and I were having dinner at the Regent Palace Hotel. A Canadian Army captain joined us at the table. During conversation, it turned out that he was from the infantry brigade that was opposing the Germans at the Falaise Gap. When they advanced they took large numbers of German prisoners who were like zombies. The huge explosions had been a 70 000 ton ammunition dump that the Allies didn't know existed. The other bonus was that a Panzer division, sheltering in the forest adjoining the dump, had been totally wiped out.

Le Havre, one of the French ports on the English Channel, was captured after a week's bombing. Eleven thousand prisoners were taken and only fifty Allied soldiers were lost. Brest was similarly attacked by bombers in support of the ground troops.

Alan Stutter tells his story of an attack on Brest, in support of Bradley's forces, on 14 August 1944 and, subsequently, of an attack on Calais, with similar objectives, on 24 September 1944:

Ever since the fall of France, Brest had been the subject of repeated attacks by all arms of the Air Force. It had been host to German pocket battleships and cruisers, U-boats and their pens and depots as well as being a major assembly point for German invasion barges. All these had required determined efforts on the part of the Air Force to destroy them.

The Germans had, naturally, installed very strong anti-aircraft defences and these had levied a very significant toll in all the raids. As a result, nomination of Brest as a target brought a hush to briefing rooms and very thoughtful looks to the faces of the participants.

After D-Day Brest figured more on the news as an Army objective and did not seem to loom large in the Air Force, which had plenty of other things to think about. It was, then, something of a shock when, on 14 August 1944, the curtain was drawn back from the briefing room map to show the track ribbon running from our station to Brest. There was an immediate air of anxiety which was not lessened by the fact that the trip was to be in daylight.

Apparently, the Army had the place surrounded but could not make much headway against the German fortifications. Our Group (5 Group) was to plaster the fortifications and let the Army in. We were assured there would be no intervention by German fighters. The German flak was still in place but this, we were assured, was unlikely to be a threat since there was a personal assurance from General Omar Bradley that his artillery was well emplaced round the city and any German gun which opened up would be quickly silenced. A sceptic

Notes used by Wing Commander Doubleday to brief crews for an operation.

might well have asked, if they were that good, why was the Army held up and asking for our help?

There didn't seem to be much point in asking the question so we set off. It was a nice, bright, clear and sunny afternoon over Brest. The only cloud in the sky was a big black one — and that was over Brest due to the smoke of bursting flak.

One could not help but wonder whether the American gunners had misunderstood their instructions. It rather looked as though they might have been trying to shoot down the flak bursts rather than the German guns.

However, there was nothing for it but to head on in, straight and level, with the bomb doors open and make sure there was no one directly above you with his open. I'd had one very good mate and his crew killed by collecting another Lancaster's load over the target and had no intention of going the same way, or at all, if possible.

We got our bombs away and had just closed the bomb doors when there was a commotion in the bomb aimer's compartment. He suddenly appeared in the hatch with a startled look on his face and blood running down his cheek. A piece of a shell had punched a hole in the plastic bubble in the nose and a piece of perspex had cut his cheek.

We soon settled down and pressed on out of the flak. However, a minute or two later the temperature gauge for the starboard outer started doing odd things. A piece of shrapnel must have punctured the radiator and the coolant was fast going. The propeller was quickly feathered and the engine switched off. This was no real problem — the Lanc flew quite well on three engines.

But it was not our lucky day. A few minutes later the process was repeated with the port outer. We were fortunate that we had the two inner engines left and the aircraft was nicely balanced. The two engines left had to be pushed a bit to keep up with the other aircraft until we got clear of France and over our own coast. Then everyone else pushed off for home and we followed at a more leisurely pace.

It was well and truly dark by the time we got back to our station. The only difficulty we now had was that there was no reserve of power if I didn't get the landing right first time. I therefore decided to make the approach a bit higher than usual as I couldn't afford to undershoot.

In the event I overdid it a bit. We touched down well up the runway with plenty of speed on. Full brakes and cutting the motors stopped us right on the end of the runway.

It was rather sad to see our poor old aircraft towed away to be repaired.

24 September 1944 was a very nasty day on the stations of No. 5 Group of Bomber Command. Cold rain slashed down from very low black clouds which raced by, looking to be only a few feet above the roofs of the hangars. It was truly one of those days when not even the birds would fly.

Aircrews were elated and confident as they looked out on this miserable scene. There would certainly be no flying and we could look forward to an early stand down. Some were already ringing their girlfriends to arrange early meets and others were planning a happy afternoon in some of the local hostelries.

There was consternation and open disbelief when fuel tankers and bomb trolleys were seen moving round the perimeter track. Then crews were summoned to flight offices for operations. Morale was not high among those who finally assembled in the briefing room.

Tension eased slightly when the target and route were revealed. There was practically no flying over enemy territory. The route marker on the big briefing map finished at the French coast. The target was Calais.

The Army was held up by stubborn German resistance in the dock area of the city. We badly needed the use of the harbour to supply our forces, so Bomber Command was asked to speed up the process of dislodging the Germans by wiping out the warehouses in which they had fortified themselves.

We were to bomb in daylight from low level in the interests of accuracy. Our own troops were close by and it was important to bomb on a precise heading to avoid hitting them while causing the Germans as much discomfort as possible. The minimum safe bombing height for the bombs we would carry was 2500 feet and this was the height from which we would bomb.

All this sounded pretty reasonable. There was an air of interest in the room. It all fell apart though when the Met Officer announced that, apart from the high winds and rain, visibility was very low all over England and the target and the cloud extended unbroken from 500 feet to 25 000 feet with severe icing at high levels. This condition was expected to continue for some time. Hilarity swept the room. The whole enterprise was plainly impossible. This brought down coals of fire from the CO

The tenor of his remarks was that the Army was in real trouble and it was our bounden duty to help them. This was an urgent job that HAD to be done TODAY and DONE IT WOULD BE! The fact that we would be unable to see or bomb the target, assuming we got there, and would be probably unable to land when we got back seemed to be of no importance whatever.

Crews headed off to get dressed and get transport to the aircraft. However, commonsense finally prevailed somewhere in the Higher Command. It was suddenly announced that the trip was on early next morning. Crews would be woken at 3 a.m. and proceed to their aircraft for take-off at first light.

Things were no better in the morning. We spent an uncomfortable day of on again off again until sent off to have an early lunch in readiness for a two o'clock take-off.

As I sat in the aircraft with the engines running, waiting for my turn to take off, the weather was obviously just as bad as the day before. The wind was screaming at right angles to the main runway so we had to use the shortest runway. I told the flight engineer to make sure he got the undercarriage up in record time as the runway looked pretty short for a load of eleven 1000-pounders and four 500-pounders. He didn't need any urging.

Our acceleration was nothing to write home about. The end of the runway was getting uncomfortably close when I hauled the stick back with the airspeed needle just hovering on the minimum flying speed. We flashed over the fence with the controls feeling soft and spongy and the aircraft feeling staggery.

At 500 feet the ground disappeared — we were in the clag. There wasn't much future trying to fly under the cloud at that height and with that visibility so I decided to keep climbing and see what might turn up. To our amazement and relief, we broke into clear air at 5000 feet. There was a layer of 1000 to 2000 feet between two impenetrable cloud banks.

We continued serenely on our way until the navigator told us we were over the south coast and we set course for Calais. I let down through the cloud, vowing that, if we were not out of it by 500 feet, we would go home. Suddenly, we were out in the clear at 1500 feet over the Channel in the middle of a gaggle of Lancasters all going the same way — thank goodness!

Calais appeared ahead. I selected the radio frequency to get the raid controller's instructions but could only pick up a light music program from the BBC. As we settled to the run in this program finished and transmission closed to the patriotic strains of 'God Save The King.' I can think of more encouraging music to accompany a charge over a hotly defended target.

Our aircraft was obviously going to be off heading when we bombed so I pulled out of the gaggle to make a complete turn over an empty looking bay just to the south, with the aim of getting the heading right on the next run. The radio was now completely silent. We were coming round quite nicely when some sportsman opened up with a Bofors gun from the shore. A stream of green and yellow tracer whipped over the middle of the aircraft. I was sure the mid-upper had caught it — but he answered when I called him and I saw him after

Halifax bomber at 466 Squadron RAAF, Driffield, Yorkshire, setting out on a mission.

we landed so I guess I was mistaken. Immediately after this someone in the crew started whistling a popular song — 'I wish I had a paper doll to call my own'. No one has ever owned up to this — but I suspect the mid-upper — it was a favourite of his.

We had now lined up and started our run. Not another aircraft in sight, except for a couple burning on the ground. We had the Germans all to ourselves. They were very generous too! They gave us everything they had. I jammed the throttles open as far as they would go and went for it. The tracer was very pretty but I was convinced that 'THIS WAS IT'.

I couldn't see how they could miss. We had to fly straight and level to drop the bombs — a sitting duck. I vividly remember to this day two streams of tracer crossing one another in the space between the two starboard engines and the leading edge of the wing. A minor error in tracking and we would have had it.

By some miracle we escaped without a scratch. I felt the bombs go and heaved the stick back and headed for the shelter of the clouds. We made it! Just to add to the fun we had bombed at well under the safe bombing height.

Of course, the weather hadn't cleared at home. We received a signal diverting us to a station in the west. We got there well after dark.

By the time we were debriefed — which took some time as it was a training station and not used to this sort of thing — and quartered for the night, the bar was closed and we could only get a very scratch meal.

I guess it was only right that the whole thing ended on the same note of disorganisation as it began.

SEPTEMBER SORTIES

While most of my comments have been obtained some fifty years after the event, by personal interview, a few excerpts from diaries prove interesting for their contemporary flavour.

Arthur Dutch's diary gives some details of his flight to Stuttgart on 12–13th September 1944:

Load the same as last night — cookie and small incendiary clusters. Off to Stuttgart — low-level right into France — had a great view of southern England and north-west France. Fun started before we reached Germany — saw one kite go in — bags of twitter[2] and searchlights — saw about five go in — one lit up the Rhine like day. Night was fairly light — saw quite a few kites buzzing around over the target. Quiet trip back to base about 2.30.

It was difficult for a navigator to maintain strict accuracy. He had Gee (a radar aid), in some cases, and H2S to give him some idea of his position but these were not infallible. Where these were not available it is little wonder that he was sometimes a little off course. If this happened, quick decisions were called for to rectify the situation. The difficulties of this can be gauged from Noel Gilmour's (466 Squadron navigator) story of his approach to an attack on Kiel on 15/16 September 1944:

We had to fly north, over the North Sea for quite a distance, at 2000 feet — below the radar cover. When we got to halfway up the Denmark Peninsular we had to turn right and climb up to operating height of 18 000 feet, swing across the northern tip of Denmark and then turn south, straight into Kiel.

At 2000 feet you're not visible to your enemy, which is very handy, but your not able to receive your own Gee either; so your sort of flying blind.

But our H2S was working that night and we turned right at the proper turning point. At the briefing they had said that the only thing we had to worry about was the island of Sylt which was crawling with flak. As we climbed up and headed towards the Danish coast, there, clearly visible on the H2S was the island of Sylt; and we were in the middle of it. We were fifty miles south of where I thought we were.

I had to make a quick decision what we should do. We could either turn left and head north to rejoin the original track or turn right, cut across the route, and join the force going in the first wave. That's what we ended up doing as you can't catch up time.

We had got far behind our scheduled time but, as it happened, so had everyone else as we were all working on the same data.

As soon as we turned on to a southerly course we were plagued by fighters who started dropping fighter flares. So we ran into the longest period of corkscrewing we ever did — just corkscrew, corkscrew and corkscrew. Our flight engineer got hopelessly sick — but we didn't get hit. So once again it struck me that if a fighter found that you were wide awake and took evasive action, he usually didn't follow through.

GENERAL TARGETS, OCTOBER 1944

Targets in the Ruhr were always heavily defended. It was commonly claimed by aircrew that you could put your wheels down and taxi over the flak on these targets. Stekrade was one such target and Ted Eagleton has this to say about his visit there on a daylight operation on 6 October 1944:

The target was very heavily defended by anti-aircraft fire. It was a delightful day — 10/10ths blue sky — not a cloud in it. Not a good day for aircrews flying over enemy targets, although a good day for bombing.

The Lancaster boys were briefed to go to Gelsenkinchen. On the way to our target there was a huge mass of anti-aircraft fire going on ahead of us. I said to the boys, 'Thank goodness we're not going there. The Lancs are getting belted today.'

No sooner had I said that than the aircraft on my starboard side got smacked. We were in formation — although not as tight as the American formations. My aircraft got the remains of his direct hit. I had five or six holes in the aircraft, including the petrol tank. But the tanks were self-sealing and that saved our bacon. We survived that and the formation still held. We saw a couple of parachutes come out of the hit aircraft but no one got home.

Then, off we went again, and we were smacked again.

Johnnie Goode, who was our other flight commander, was on my port side and said, when we got back to base, 'You'd taken us through enough. We almost decided to let you go on your own.'

We dropped our bombs, but I'm not sure how successfully. It was the hottest flak barrage that I ever flew through.

One hazard, on return to base, was fog. Too often, aircraft which survived a raid crashed when they couldn't find somewhere to land.

2. Twitter — RAF slang indicating twittering (or contracting) of the sphincter muscles due to fear or apprehension.

3. F.I.D.O. removed the fog from the area of a landing strip by heating the air and evaporating the water particles of which the fog was composed. This was done by placing large burners at short intervals all round the strip and supplying them with fuel by large pipes which ran around the edges. The amount of fuel used was enormous but was justified if it would save one or two hundred aircraft.

FIDO (Fog Intensive Dispersal Of) clearing fog at an airfield.

The introduction of FIDO (Fog, Intensive, Dispersal of)[3] on a limited number of aerodromes gave aircraft a useful landing facility.

Ted Eagleton explains the FIDO facility's advantage and Harry Brabin gives an example of a diversion without FIDO on return from Kleve on 7th October 1944:

I was diverted, when returning on one or two occasions, to the Carnaby strip. It had fog dispersal apparatus and could put a lot of aircraft down.

If it was only going to be used for one or two aircraft it was hardly worth operating FIDO because it was so expensive. It needed huge amounts of kerosene fuel to clear a landing strip. It used to break up the fog to about 200 or 250 feet. You could then see the lighting system on the strip.

It was not easy to handle. It was quite turbulent coming in over the end of the field, but the strip was so wide and so long that there was not much danger of going off the runway.

Harry Brabin:

When we arrived back over Pocklington there were so many planes milling around in the fog that we were diverted to an American Flying Fortress base. We had been hit several times in the engines and, as we circled, we lost power and landed in one of the fields between the runways. We had trouble in stopping but eventually pulled up a few feet short of the control tower. We could see the American controllers watching our propellers revolving right before their eyes.

We treated ourselves to bacon and eggs in the PX store and then a hefty meal in the mess. Just as we were about to leave an officer brought each of us a huge T-bone steak — 'With the C.O's compliments gentlemen.' How could we refuse? One of those steaks would have fed the whole crew at Pocklington.

That night we slept on portable beds in a hangar. We could hear one another's groans as our tummies tried to cope with this unaccustomed feasting. I for one was unable to cope. What a waste!

As we were taking off, Bill thought he would show off by taking as short a run as possible. American aircraft needed twice as much distance as ours for take-off. Bill applied the brakes at the start of the runway, ran the engines up to maximum power, released the brakes and off we went. We often wondered what the Yanks thought of it.

Landing is always a critical time for an aircraft. Close to the ground, at low speed and low power, it is vulnerable to any miscalculation by the pilot.

Keith Burns tells of his experiences when his pilot was undergoing a familiarisation flight on his squadron in October 1944:
The Halifax MarkIII MH-T was a brand new but problem aircraft. The flight commander had, three times, set out for Germany in it and had turned back each time because of sudden overheating of the starboard inner engine.

We were the new crew on the squadron and were detailed for a 'bullseye' so the flight commander decided we should try out his new aircraft which had been rechecked for the ninth time. He took it up in the morning, with us along for the ride, and did everything with it that one can do with a Halibag, returning satisfied that all was right.

The exercise consisted of a cross-country which finished at the bridge over the Bristol Channel, where an infrared light traced a path on infrared film to indicate the path of our 'bombing' attack. Just as we turned out of the 'target' the flight engineer called up that the starboard inner engine temperature had gone through the roof. Now, the Gee, in a Halifax, was powered from the starboard inner, so I called out to give me half a minute and then they could do what they liked. I wanted one check fix that we were on the right track for home. From there I proceeded to map-read my way home and, as it was a good night, there was no problem.

However, another problem existed. The skipper was very worried that he had never landed a MarkIII on three engines. The flight engineer tried his best to reassure him, pointing out that he (the skipper) had several times landed MarkII and MarkV aircraft — but to no avail.

We levelled out too high and touched down two-thirds of the way down the runway. We went through the boundary fence, across the road, through a tall hedge, across a deep agricultural drain, which took off the undercarriage, and finished with the nose in the mud of a field of sugar beet.

We climbed out of the wreck and went across the ditch, through the hedge and on to the road. By this time an ambulance had arrived, manned by a Scottish corporal about five feet tall. He then tried to find seven airmen in one place at one time and could not get past six at any count — repeatedly asking, 'Where's the other mon?' I think the 'other mon' had gone back to the aircraft to retrieve his flying ration of York chocolate.

It finally got too much for my laconic, Cairns-born, rear gunner. He pulled his Thermos flask out of his flying boot and said to the diminutive Scotsman, 'ere mate. 'ave a cup of coffee. Steady your nerves.'

The phrase 'press on regardless' was used, mostly in cynical jest, by aircrew when there was a difficulty which could not be avoided and to which there was no immediately apparent solution — such as a highly objectionable target which had to be attacked. Stories of how crews 'pressed on regardless' were commonplace at debriefings. It was also used as a 'given name' for some keen (and over keen) squadron and station commanders who insisted on crews operating despite what appeared justifiable reasons for aborting operations.

To be fair, the burden of command required commanders to be 'keen types' to ensure maximum effort was available for operations.

Several examples of this have been given to me and the first was provided by Arthur Dutch's entry for 19/20 October 1944, when he attacked Nuremberg:
Out to 'Charlie' (aircraft) about 4.30 p.m. Just going up on air test. 'Charlie' landed 5-ish — had to be fuelled and bombed up. No. 3 pulsometer pump not working. Thought we'd be scrubbed but W/Cdr says 'go' — means about 114 gals of juice we can't get at. Took off about 6 p.m. — the rest had 20 mins start but we caught 'em by the time we reached the target — Nuremberg. Another attack on Stuttgart as we went in — saw it plainly; only about 20 miles off. VHF didn't work. Trip back seemed long — only about 5 hours, one of last back.

Doug Glasson, on his second tour with 466 Squadron RAAF had a similar experience:
The Gee was essential to sensible navigation on operations.

I well remember on one occasion we were going to a tough target with a very intricate series of legs to be negotiated. We air tested but when we went out to the aircraft that evening the Gee wouldn't work. Not a blink out of it. The radio was a bit dickie but there was absolutely no Gee. So I had the CO summoned to report it

He was great on maximum effort. If you could get into the air you were on. This was a maximum effort and a little problem like Gee didn't matter. I pointed out to him the intricacy of the navigation involved in this operation and said that my view was we should not be on it.

He walked round the aircraft three times and then he came over and said, 'You've got five minutes to take off.' The rest had already gone.

I subsequently talked with that CO about the episode and he remembered it distinctly. I said, 'Now, looking back on it, maximum effort or no, I wonder, if you had the opportunity to take the decision again, what would you do?' He said, 'I'd say stay on the ground.' It was a very critical time of the war.

Arthur Doubleday tells of a similar incident — from a different point of view:

As squadron commander, I went out to see the planes off. At one dispersal I came across a crew sitting outside a plane. They said they couldn't go because their pilot had violent stomach pains. I grabbed him and rushed down to the medical officer, who was watching the take off as usual.

I said, 'You've got five minutes to examine him because it's ten minutes to last take-off. He's either got to be certified not fit to go or he's got to go!'

The doctor was back in five minutes. The chap flew and came back full of enthusiasm. He'd had a wonderful trip, and after he'd had his bacon and eggs, went off to bed in high spirits. So I rang the MO and said, 'That was marvellous — what did you do for him?' He said, 'I didn't do anything for him. I just gave him an Aspro.'

Arthur Dutch, in a diary entry dated 31st October 1944, recalls being hit by flak:

We were deputy leader. I had bags of fun pumping off green Vereys for the boys to form up on us. We went into the Ruhr — to Duisberg. We couldn't bomb — no markers down.

The winds were just coming through when we were hit — on the way to the target. I felt the jar — the first I knew was Jack putting on his chute. I promptly came on the intercom — couldn't hear a thing — someone had left his switch on thus jamming all our intercom. Really scared — they were really throwing the flak up — nearly everyone was hit. Had a hole in the windscreen — the perspex hit Al — a few in the leading edge and a few in the tail.

Landed No. 2 — the other deputy leader beat us in, but only just. Everyone feeling great, the tour down to thirty. Eight crews on the station finished — four from each squadron. Boys out to celebrate.

ASSISTANCE IN THE WAR AT SEA

From time to time Bomber Command attacked targets which were of significant importance to the war at sea.

To restrict U-boat activities, an attack was launched on the U-boat pens at Bergen on 2 October 1944 and Trondheim on 22/23 October 1944. Only forty-seven planes bombed at Bergen as cloud obscured the target and the Master Bomber called the raid off. Wireless Operator Arthur Dutch's aircraft was one of those which bombed at Bergen and, later, at Trondheim. His diary extracts tell the stories:

Surprise of my life when I learnt we were going to Bergen in Norway. Bags of special gen for us. Took off about 10.30 pm after spending a couple of hours sleeping. Trip was quiet all the way. Jack [the navigator] had Gee all the way — thank goodness. My help would have

been negligible: the broadcast stations too weak for loops [direction finding loop aerial which allowed the wireless operator to get bearings]. First run over the target couldn't see anything for cloud (entire trip through cloud). Came back just missing a couple of Lancs in the stream. Bombs went OK Flak inaccurate but plenty of it. Diverted on the way back. Bod who sent the diversion made a horrible boob — sent the code version and then the whole lot in plain language [This would have compromised the Bomber Command code]. In cu-nim most of the time — had to earth everything. Sparks a couple of inches long every time I attempted to unearth.

Arthur Dutch made these notes about the trip to Trondheim in his diary:

Quite a trip — Trondheim U-pens under construction — have to pass through a front down to 600 feet; supposedly only 100 miles long — Took off about 3.45 — got into cloud just after crossing the English coast about Hull — didn't break cloud until 100 miles from Norway. Looked pretty in the moonlight — snow covered everything, including the icebergs that glistened in the fiords. Not much flak, no fighters. PFF couldn't mark — Hun beat us with a smokescreen [master bomber ordered everyone to abort] — everyone could see the town clearly. Elevators stiffened up — Harry could hardly push 'em. Into cloud again about the same place — didn't get out till England — had only just enough juice, nearly landed at Coventry. I worked quite hard — loops when out of Gee range, with bags of messages from Group.

*Les Mills, an electrician on the ground staff, was stationed at Sumburgh Head in the Shetland Islands. A force of Lancasters took off from there to raid the **Tirpitz** in Trondheim Fiord — and severely damaged her. They were accompanied by a photographic aircraft. This was a normal Lancaster which carried movie cameras and photographers, but no bombs:*

When the Lancasters took off from Sumburgh Head in the Shetlands to bomb the *Tirpitz* at Trondheim, they were accompanied by a photographic aircraft. All the bombing aircraft returned safely but the photographic aircraft was quite late getting back.

It was quite badly damaged, probably because it usually made a couple of circuits of the target after everyone else had left to try and get some good shots of any damage.

Anyway they were in some trouble. They were not at all sure they could make it back and got to the stage of jettisoning loose equipment to try and stay airborne.

One of their problems was that the tail had been damaged and the rudder pedals were ineffective. In the course of throwing things out they came across a piece of rope and found they could rig this to give some rudder control and help get them to Sumburgh Head.

OPERATIONS, NOVEMBER 1944 TO FEBRUARY 1945

Essen was one of the major Ruhr centres and the home of the Krupps armament works which were damaged in a raid on 28/29 November 1944. There were no reported losses but I doubt that Jim Tudberry would agree with this. He had this to say:

We were going to Essen and just past the point of no return when the starboard outer engine packed up. We couldn't feather it so we carried on to the target area. The prop kept windmilling all the time, the engine caught fire and we put it out with the extinguisher.

We got up to 16 500 feet, which was the bombing height, and the port inner engine was going pretty rough and not fully powered. As soon as we opened the bomb doors we dropped to 12 500.

We bombed from 12 500. On the way home it didn't look too good. The starboard outer was still catching fire and we kept putting it out but we couldn't feather it so we carried on in a straight line for Manston in the south of England, losing height all the time.

We were over the Pas de Calais area at about 750 to 500 feet when the aircraft started vibrating so badly I couldn't read the instruments. So I decided to make an immediate forced landing. I landed safely in a field, which I found had been demined and had the antiglider poles taken down only three days before. All the crew got out safely. It was dark — about seven o'clock in the morning in November.

We were picked up by the Army and taken to Boulogne. There the Air Force acquired us and an MTB took us across the Channel.

Some German towns escaped a visit from Bomber Command until late in the war. Heilbronn was such a town. It had only one visitation — on 4/5 December 1944.

Arthur Dutch's diary records the visit:
Going to Heilbronn, simultaneous attack on Karlsruhe. Taking a second dickie. Quiet trip to target. Target well marked; well alight as we let them go; cookie and incendiaries (cookie well labelled). Caught a glimpse of a jet assisted fighter. Attacked by a Me-210, corkscrewed like the clappers. He didn't fire. Ken got a few hits, then three guns jammed. Johnny got about 500 rounds away as he broke away up — got a fair number of strikes; attacked again. Ken had another go — missed, he thinks. Zoomed into cloud. Phew! cloud turned out to be a cu-nim — iced up — had to turn back — went north almost to Karlsruhe — it was burning beautifully and we went around it. Got sparks about three inches long from the trailing aerial in cloud.

John Thomas comments on a trip to the Ruhr:
On 24 December 1944 our target was Mulheim airfield. About twelve major German airfields were being bombed by the RAF and US Air Force on the same day. Our bombload was a mixture of HE with some 250-lb anti-personnel stick bombs. The latter would explode some feet above the ground and act as daisy-cutters to damage enemy aircraft.

When we reached seventy miles an hour on our take-off run the needle of the airspeed indicator started to shudder and then fall back as though we were losing speed. However, we were still accelerating with obvious full power. I had to decide in a split second whether to carry on or cut the throttles; with the prospect of going through the fence at the end of the runway and perhaps having to raise the undercarriage and drop it on its belly. The big worry was that we had been briefed that the stick bombs were very sensitive to any impact or sharp deceleration.

I was reasonably sure that we could get off and fly safely, so we 'pressed on regardless' and lifted off (with a sigh of relief and an ashen-faced Jack White beside me).

Once airborne we climbed slowly and at high airspeed as I didn't altogether trust the air-speed indicator. We eventually reached our planned height and got to the target — the air-speed indicator and bomb sight were now working.

By this time I had moved into close formation with another Halifax who was showing clear signs of nervousness at the intensity of the flak. I told the bomb aimer, 'I will count 4-3-2-1-go! and on the word "go" you release.' This procedure resulted in us obtaining an aiming point photograph. We achieved 91.5 per cent aiming points on our tour.

On returning to Pocklington we found the short runway was in use and ground mist was starting to obscure the far end. I was not prepared to risk a landing without the ASI so we diverted to Carnaby. They lit the FIDO after we landed and we watched a US squadron of B-17s land.

As we walked along the line of Halifaxes on our way to debriefing we came to one with its nose missing. A bomb from an aircraft at a higher altitude had smashed the nose off and removed the bomb aimer. Most of his parachute harness was still hooked in the wreckage. The Canadian pilot was running round in circles saying, 'I'm LMF! I'm LMF!'[4]

We travelled back to Pocklington by bus.

4. LMF — lacking moral fibre. This was the expression used by the services to describe cowardly behaviour. The penalty for being declared LMF was imprisonment, immediate loss of all rank and likely dishonourable discharge.

B-17 Flying Fortress.

(Dennis Hornsey Collection)

1 JANUARY TO 28 FEBRUARY 1945
GENERAL TARGETS

John Thomas had a close encounter with a Fw 190, as he relates:

On the night of 2 January 1945, we bombed Ludwigshaven — a twin city with Mannheim. To get there we flew a south-easterly course which took us south of Stuttgart when we turned to the north. This took us close to the western edge of Stuttgart and we drew quite an amount of flak from the city defences.

It was only some two or three minutes after turning on to the northerly course that my windscreen was completely filled by the silhouette of a black FW-190, hood open and the pilot clearly visible in the glow from his own exhaust. He was about twenty-five yards in front of us and passed from starboard to port. It was obvious he hadn't seen us.

I checked with Intelligence the next day and was told that FW-190 fighters flew with the canopies open at night as the reflection of the exhaust from the inside of a closed canopy obscured any external vision. It must have been freezing in mid-winter.

Command or Group, in some instances, devised methods to increase chances of survival. Ross Pearson, a wireless operator/air gunner of 102 RAF Squadron (many Australians served on RAF squadrons) tells of one not-so-successful ploy:

I can remember an attack on Magdeburg on 16 January 1945. As we crossed the North Sea we had to climb to avoid cu-nim cloud and so were picked up early by German radar. In consequence we were losing planes on the way across the North Sea. Before we reached the enemy coast I had seen two planes blow up — I suspect one was from my own Squadron — No 102.

Then the long haul in to Magdeburg. All the way we could see the occasional plane go down.

We bombed and turned for home. We had been briefed that, on this mission, we should dive regularly to a different level on the way home in order to confuse the fighters. This we did , but at every level the Germans were a step ahead of us. There to meet us were rows of flares, like street lights, to light up our path.

Pilots also developed their own individual tactics. Bruce Otton, 466 Squadron RAAF, summed up his approach as 'Flying high':

My view was that we had a job to do. I certainly didn't want to be heroic and neither did any of the older members of the crew. We wanted to get the job done and get back.

To that end, because I had seen on daylights that some squadrons tended to fly high above the stream and drop their bombs through other aircraft, I made a point of being at the top.

To help navigators in the main force, a system was set up whereby each squadron nominated one or two aircraft on each sortie as 'windfinders'. Each of these aircraft transmitted, at intervals, to Group HQ the wind it had determined. The values were collated and used to establish a 'standard' wind to be used by the whole force for the next time period.

Noel Gilmour, a navigator with 466 Squadron RAAF, flew on a raid to Bohlen on 13/14th February 1945 and reports on his experience with windfinders:

We were on a trip to Bohlen, just south of Leipzig. There were predictions of westerlies which would slow us down on the way home. In fact these winds were eighty miles per hour and, once we'd gone into Bohlen and headed back in a westerly direction, we were in trouble.

We headed back with our H2S sometimes working and sometimes not. When it worked well it was great but quite useless when it didn't. We corrected our position on D/R and using the wireless operator's half-hour Group broadcast. Group would give us a wind that was averaged out from those received from wind finding aircraft.

Every squadron had one or two aircraft that would be a windfinder. They had to have a reasonably competent navigator who was lucky enough to be right most of the time. They would be nominated windfinder for the night.

On these occasions the W/Op would break radio silence and transmit to Group, in code, the wind the navigator had found. From these Command would establish an 'average' wind for areas. Every half-hour or so there would be a Bomber Command broadcast giving a wind.

The whole aim was to keep the stream together. With each navigator using his own estimate of the wind, the stream quickly became scattered over half of Germany. Even if they all started together from one point, they would inevitably go off in different directions.

This could be minimised if everyone used the same wind. If a navigator could get a good fix as well the whole thing should work like clockwork. It undoubtedly helped someone in trouble but that was not the main aim. While navigators were free to ignore the command wind and use their own, that increased the risk of separating his aircraft from the main body and increasing the risk of being singled out for attack.

We were bucketed along at a rate of knots. It wasn't rough but there was a lot of thunderstorm activity. It was the first time we had St Elmo's fire on our propellers. Then the aircraft started making funny moaning noises. We were building up a big load of ice. We couldn't go up so we went down to a warmer level to get the ice off.

As a windfinder for that night I was finding the wind and sending it back. I got a very lucky Gee fix near the Ruhr — slap bang in the middle of Frankfurt-on-Main. I had just done it when we were jumped by a fighter — only one luckily.   There was an exchange of gunfire and corkscrewing. After we'd corkscrewed, dived and fired a few bursts he broke off. From then on we just cruised on across France.

Sam Weller was on the same operation and had the following experience:

The weather was bad. It was one of those trips in the northern winter. We were due to start about four o'clock in the afternoon but the flight was put back until midnight.

We finally got on board the aircraft and taxied out. It was pouring with rain. The aircraft which took off in front of me stalled, turned over, crashed and burnt.

Ross Pearson (wireless air gunner), John Thomas (pilot) and members of crew and ground staff, 102 Squadron RAAF, Pocklington, Yorkshire, 1944. Front row, from left: Shorty Hughes (flight engineer), Ross Pearson. Second Row: Jack White (bomb aimer), Nat Goldberg (air gunner), John Thomas, Jack Williamson (air gunner), Derrick Turner (navigator). English ground staff unidentified.

I, by mistake, taxied off the perimeter track. Fortunately the plane didn't stop, or we would never have got it out of the mud. With throttles full open I kept it moving and got it back on the track. By now I was nearly a nervous wreck.

This was a trip where nothing went right. As we started down the runway the airspeed started to build up and then went back to zero. By this time we were too far along the runway to stop and I had no option but to take off.

I had to fly this thing with no airspeed indicator. I could only watch the artificial horizon and hope I got it right.

We crawled up and eventually broke out of cloud at about 24 000 feet. The navigator said, 'How am I to navigate with no airspeed indicator?' I said, 'This is a good time to learn because I'm not going back down through that cloud now.' Going down with no airspeed indicator wasn't my idea of fun. 'Sorry, but you'll just have to work out some sort of airspeed.'

About forty minutes to an hour before we got to Leipzig there was a hell of a row from the number four engine and it blew to pieces.

We were up about 21 000 feet and I simply had to come down. But, with no airspeed indicator, I couldn't tell how close we were to a stall. Every time I tried to level out the plane would stall and I would go down further. I finally levelled out at 11 000 feet.

The navigator asked if I was going to turn back. I said, 'No. We're nearly there.' We weren't, but I wasn't going to turn round and risk running into the other aircraft.

We got to the target, dropped our bombs, and the bomb doors wouldn't close — hydraulic failure. We had a chap called Heath flying with us. He was a Canadian navigation instructor who had come for some experience — he was going to get some.

I said to the engineer, 'You've got to get those doors closed. I can't fly with them open and no airspeed indicator.' So off he went and Heath went to give him a hand. There was a little row of lights which showed when the bomb doors were open. I watched them but nothing happened.

Eventually they came back and said, 'We can't get them closed.' I said, ' I've got a great incentive for you. You either get them closed or we bale out. We can't get back like this — we won't have enough fuel.'

They went back. After what seemed about four hours, but was probably four minutes, the lights went out.

So I pressed on.

After some time the navigator said, 'I haven't the slightest idea where we are. The Gee's not working, the H2S isn't working, nothing's working. (They all worked off an alternator in the engine which had failed.) I can't tell you where we are.'

I said, 'I can tell you.' He said, 'How can you tell? Don't be funny.' I said, 'I can tell you exactly where we are.' He was down in his dark hole. I said, 'Come up here.'

He came up. It was daylight by now and we were running along an island off the Dutch coast. Beautiful for navigation.

'What are you going to do?' 'You can take a fix when we get out over the water. I'm going to stay on top of the cloud so that we can dive into it if anything comes up. Because we're so short of fuel and uncertain of our position I'll send an SOS for a fighter escort when we get across.'

'Don't be funny! The wireless hasn't been working for hours.' chimed in the wireless operator.

There we were — on our own with nothing working. So I just sat on top of the cloud and went over the North Sea to England. When we got there and came under the cloud we were almost right over our station.

Dresden was attacked on the 13th February. The city was crowded with refugees and any attack had to have disastrous consequences for civilians. After the war, this attack provided a focus for criticism which tended to obscure the great efforts of Bomber Command.

Dudley Hannaford relates his experiences:

We were attacking a target fifty or sixty miles south-west of Dresden on 14 February 1945. We were over the target and the pilot (Padge) said, 'Well, it's just a piece of cake, boys.' I'd given the wind to the bomb aimer and the engineer had gone down to help push out the window. All of a sudden the plane lit up. It was as bright as midday. We were caught in a cone of searchlights — which was followed by flak which hit just behind the Cookie. Fire broke out in the fuselage.

The mid-upper gunner was having great difficulty battling with the fire, but everyone reported okay. Then the rear gunner called out 'Again'. The pilot took evasive action and, the next thing, (so he told me later) the two port engines were hit and fell out. The plane became uncontrollable.

The rear gunner was in trouble. He couldn't rotate his turret as the hydraulic drive came from one of the port engines. I tried to go to his help but I couldn't get through the flames. I believe the mid-upper also tried to get to him and couldn't. Then the plane went into a stall — followed by a diving spin.

I fell on top of the navigator as we tried to escape. The rear gunner reported that he'd turned his turret manually and was leaving. The mid-upper got out through the side exit — his trousers and parachute were on fire. As he went out he was trying to put out the flames in his shroud lines. He got down all right. . . but with scarcely any trousers left.

The navigator and I were huddled together down towards the bomb aimer's compartment. The centrifugal force held us there. Then, after some time, the plane decided to spin in the opposite direction and I said to the navigator, 'Go!' The escape hatch had been opened, we presume by the bomb aimer, so the navigator left by it and I followed him.

I saw the pilot leave the controls. He told me later that it was only 500 feet when he jumped from the plane. I was surprised he was able to get out unscathed.

Chemnitz was raided on 14/15 February 1945.

The Australian Halifax squadron No. 466 sent sixteen planes on this sortie. The aim was to disrupt communications to the Russian Front. Chemnitz was an important traffic centre.

F/O Bruce Otton of 466 took part:
This was our first trip. We survived it in spite of ourselves.

We climbed to some 6 000 feet to cross the coast and then climbed to operating height.

Suddenly the rear gunner came up and sat beside me. I said, 'What the so and so are you doing here?' He said, 'My oxygen's not working.' So we were weaving practically all the way to Chemnitz because we didn't have a rear gunner. I later found out he had checked his oxygen on the ground and it hadn't been working then. He never did that again.

I didn't turn back because, quite frankly, I believed that you could get away with it if you didn't have a gunner in the rear turret. I warned the rest of the crew to keep their eyes open and had him glued to the windscreen on the starboard side to see what he could see. I felt we could get by.

These precautions were coupled with judicious weaving.

CROSSING THE RHINE

*In **Bomber Offensive** (p. 172), Sir Arthur Harris wrote that the crossing of the Rhine was one of a series of bomber attacks which helped the army achieve its objectives without vast sacrifices of manpower and resources.*

One of the Rhine towns so attacked, or marked for attack on this occasion, was Wesel, visited by Fred Wright on 17 February 1945. he tells of his experiences:

We went to Wesel. It was 10/10 cloud all the way and the target was completely covered when we got there. The master bomber told us to come down from the 20 000 feet we were at to 5000 feet. But it was no better there. So he scrubbed it.

We were still in cloud all the way back and our base was blanketed when we got there so we were diverted to Mildenhall.

When we were debriefed we were asked if we had brought our bombs back. We had, so they rushed out and moved the aircraft away to a more isolated spot. Next morning we were directed to fly out to sea and drop them in the Channel. We were not very welcome.

OIL TARGETS

*Hilary St George Saunders, in **The Royal Air Force, 1939-1945**, says: 'The really decisive blows against oil and communications fell towards the end of the war' (Vol. 3, p. 387).*

A successful raid on the oil production centre at Reisholz, near Dusseldorf, on 20/21 February 1945 halted oil production completely in the area. Bruce Otton was on this raid:
On my first trip we went to Chemnitz, in the Ruhr on DR. When we got back we were the heroes of 4 Group. We were only four miles off course, whereas the rest of the force was off by nineteen miles. My navigator was a hero. The squadron commander thought it was marvellous. The navigation leader came up and said, 'You've got a marvellous navigator there.'

Consequently, on our second trip, I felt 'I'm in good hands.' and away we went. We took off in late afternoon.

We had to turn at Reading to go out over Beachy Head. I could see Reading coming up and said, 'When will we be at Reading?' The navigator said, 'Be about five minutes, Skip.'

As I saw the other aircraft starting to turn and we were right over the top of Reading I said, 'When are we going to get to Reading?' 'We'll be there in about a minute or so.' I said, 'Right O.'

We overshot Reading. Everyone else had turned when we turned and I noticed the rest of the force fading away in the distance with the navigation lights, which were to be kept on until we reached Beachy Head, rapidly disappearing. So I said 'Everything okay?' The navigator came up and said, 'The wind is 180 deg. different to the forecast. You'll have to put on some herbs.' So I had to increase power to catch up to the pack.

They used to build doglegs into the flight path so that you could, if running late, cut across it and pick up some time. However, as we were getting near Beachy Head he said, 'Let's cut a corner and climb to 7000 feet.'

Ultimately, we found ourselves coming into the target from the other end to the rest of the force. This wasn't exactly fun. Then, as we were going through the target, the bomb aimer said, 'Dive port!' in a very calm voice. I dived to port and, suddenly, a Ju-88 went straight up in front of us. My gunners fired in all directions.

It transpired that the bomb aimer was about to press the tit when he suddenly found himself about to shake hands with a Ju-88 pilot who had come into his bomb-sight. He could see him sitting in his cockpit. Calmly, with no panic, he said, 'Dive port!' and we turned away as the German kept coming up.

We got our load away and eventually got back. Next morning I was called up by the station commander. He ripped strips out of me for going to 7000 feet when I was ordered to keep under 6000 feet so as to keep under the enemy radar and keep the force concealed for as long as possible.

He also said I should be sure the navigator had a sextant on board to take astro shots periodically.

OPERATIONS, MARCH TO MAY 1945

Late in the war, late 1944/early 1945, the Germans introduced a new tactic. Fighters were used to attack aircraft which were at the end of their missions, often on the home circuit. This short-lived tactic was remarkably effective.

Keith Burns and Bruce Otton both tell of intruder attacks on the night of 3/4 March 1945.

Keith Burns:

We were flying back to base when I saw a lot of activity over a town. I checked where we were and it was Hull. There seemed to be Jerry fighters everywhere.

We got into the circuit and then, over the R/T, came the call 'Bandits, Bandits, steer to your diversion station.'

Our diversion was Gaden and there we went — across Lincolnshire. There were Lancs everywhere going like bats out of hell and we were right among them. But we finally got down.

Because we had not returned to base we were shown as missing on the list at 4 Group HQ. This gave my new wife, who was on duty there, a very nasty feeling.

On that same night, a friend of mine — Joe Moss — was actually in the funnel, preparing to land, when he got the diversion call. 'Go to buggery!' was his

Focke-Wulf 190 (Royal Air Force Museum, Hendon).

mental and verbal reaction. He just bounced his plane down on the runway in the dark.

As he pulled up at dispersal, a WAAF officer came over in a truck with her lights full on and yelled out 'Don't you know there are Jerry fighters about?' To which Joe replied, 'Turn your lights out!' As he did so, a Ju-88 streaked across the drome and they all dived under the truck. Luckily, there were no casualties.

Tired and weary, it was natural to relax when you neared home. But this changed in March 1945 — as Bruce Otton of 466 Squadron RAAF relates:

On 3 March 1945, as we came back, I said 'Look up there!' and there was an Me-109 trailing the aircraft above and not shooting at us.

We were coming up through the clouds when we noticed a pink glow in the clouds. 'Crikey!' I said, 'A couple of fellows have collided.' So we went on a bit further and I heard the call, 'Bandits in the area!'

This was a new experience for us so we called up the W/Op to see what action we should take. It was to make sure the I.F.F. [Identification Friend or Foe, a device to provide a radar indication that the aircraft was friendly] was operating. We could see some tracer flying around. As we were coming near to base I called up for a position on the circuit and I was told I'd have to hold — there were others ahead of me.(In fact I had been jumped for position by a great mate of mine who I knew was behind me. 'You stinker.' I said.)

We saw one aircraft attacked and set on fire as he was approaching base. We circled and I heard my mate call up to say he had been attacked.

We were told all lights were going out. We were to go somewhere else. It was recommended that we go north — so we did. We were being fired on by our own AA and that didn't thrill me too much so I said to the navigator, 'Give me a course for Liverpool!' and off we went.

A little while later, the engineer said, 'Hey! We've only got ten minutes fuel left.' So we couldn't get to Liverpool. So I said, 'Fair enough. I'll head south and I'm sure we will find some place to land between the coast and the Pennine chain.' So I headed south, but practically every station had its lights out. So I said, 'Prepare to bale out if we can't do any good.'

Then we came across a drome with its outer rim lights on. I checked the book and saw it was Leeming with the code letters LM. So I called up and asked for permission to land but they said, 'No! We're under attack. We can't put our lights on. You should go somewhere else.' At this stage the engineer said, 'Only a few minutes fuel left.'

So I put out a MAYDAY call and they said okay.

Call up when your in the funnel and we'll put the runway lights on and you can come in.' At that stage we were at the end of the downwind leg at about four and a half thousand feet. I said to the bomb aimer, 'We'll do a circuit and let down.' The engineer said, 'Pigs! You can and must get down immediately.'

We cut everything and went down like a rocket. As we swung into the funnels I called for the lights. Five fighter flares went right across the front of us. But we were committed and we didn't have any fuel reserves so we touched down. As we came in I saw a FW-190 and a couple of Me-110's streaking past. In fact, they had hit a hangar full of incendiaries.

We touched down and, as soon as we had done so, they switched off the runway lights and I bored into the darkness. A hero came out in a truck while the drome was being shot up and led us off the runway and into a dispersal.

I started to go through the dispersal cockpit check when the engineer said, 'You silly bastard, cut the switches and let's get out of here.' I was the last one out.

They picked us up and took us to a big sleeping quarters, where they said, 'Find yourself a bunk and fit yourselves in.'

The next morning, we went to the end of the runway, revved her up, took her to 1000 feet, did a split-arse turn, came back and shot the place up.

Manoeuvring a fully loaded four-engined aircraft on the ground in an English winter provided some unexpected difficulties. There were no problems if you stayed on the runways and perimeter tracks but trouble arose if, by some mischance, you strayed off.

Fred Wright found this out on a trip to Hamburg on 8/9 March 1945:

Our start was a bit different to the normal. As I taxied out to take off, one wheel went off the perimeter track and we were bogged. We had to shut down the engines and sit there until everyone else had left. Then a petrol tanker pulled us out. We restarted the engines, did our checks again and we got to the end of the runway just after the last plane off had set course.

We got the green light and took off, did a steep turn, came back over the centre of the field and set course. We cut the first leg to try and pick up the mainstream as they crossed the English coast.

The cloud was down to sea level and our briefing had instructed us not to climb to 10 000 feet until we crossed the enemy coast. We were to hold that height until later, when we would climb to 21 000 feet.

However, when we climbed above the cloud the enemy fighters were waiting, despite our low level approach under the radar. They were sitting there waiting for us and followed the whole way to the target. Luckily, we were not attacked.

We later found that a number of pilots had become disoriented while flying across the North Sea on instruments while in the cloud, and had flown into the water. Others, not liking the situation, had climbed to 10 000 feet immediately on crossing the English coast and given the game away.

A rear gunner, guarding the aircraft from attack, could find the effect of searchlight coning quite disconcerting. Keith Spain describes the reaction of one gunner:

I can recall being coned over Potsdam (14/15 April 1945) and as soon as we were coned the AA got very heavy. The skipper put the plane into a dive and we came out of it okay. But the funny part about it was that, as we were flying into the target, we could see all the searchlights and our rear gunner started to sing, 'I belong to Glasgow'. The skipper, who was very quietly spoken, said, 'Jock! Be quiet! We're getting over the target.'

But he continued to sing and our bomb aimer yelled out, 'Shut up! You Scotch bastard!' There was deathly silence.

MINING OPERATIONS

Mining — or 'gardening' — was covered in Chapter 4 as it involved the twin-engined aircraft of Bomber Command.

Mining operations continued until the end of the war with the four-engined aircraft playing an important role. Mines were laid around Norway, around the Danish islands, at the mouth of the Elbe, in the Baltic and the Kiel Canal. They disrupted sea communications and posed a serious threat to U-boats going on or returning from patrol.

The Germans admit to 842 vessels being sunk in areas where aerial mines were laid.

Fred Wright tells of his experiences in late 1944:

Halifaxes were often used for mining operations. We went on two gardening trips to Stettin Bay — quite a long trip as opposed to the bombing trips.

The thing that hit me first was the mine itself. It was very long and cylindrical, with a parachute at the tail end. It was so large the bomb bay doors could not be closed on it. Once it was in the bomb bay, the doors were closed until they just touched its sides. Part of it was in the open air all the time. It didn't seem to have any effect on the handling of the plane.

We flew across the Scottish coast and North Sea at about 500 feet. At the Norwegian coast we climbed to 10 000 feet. Once across, we turned south to the Kattegat and bombed on the navigator's co ordinates. We used Gee and H2S to find the spot. We opened the bomb

doors, the bomb aimer said 'Now!' and down it went.

On one trip, we were flying on and on and we could see a bit of a scuffle going on ahead. There seemed to be a lot of machine-gun fire going back and forth but it never seemed to get any nearer. We went on turned west over Norway and left the lights behind.

When I got back I found they had been the Northern Lights. It was an eerie feeling to see these lights pointing up in the air. Flying towards them, the worrying thing was that anything behind us would see us as clear as day.

Jim Tudberry:

Mining trips were, generally, put on when the weather was grim. Take-off always seemed to be in crook weather. I remember we went to Kiel and the Skaggerak at 2000 feet — icing all the way. We were still at 2000 feet when we dropped our mines — 4 x 2000-lb. mines in each aircraft, and we stayed under 2000 feet all the way home.

We never liked bringing them back. We iced up to such a degree, one night, that the mine would not release. I put the aircraft into a steep dive but it still wouldn't go. I heaved back in the stick and it released, but in doing this I cut the two starboard engines. The aircraft flipped over on her back and did a stall turn. I was lucky to regain control.

Max McVicar

We used to set up a heading on the H2S and follow it. At the time of releasing the mine, a camera automatically photographed the H2S screen.

GENERAL OBSERVATIONS

Weather

Sir Arthur Harris has often been criticised for his area bombing campaigns. But it must be remembered that the first directive to bomb enemy cities, as opposed to purely military targets, was given by Air Staff before Harris succeeded Sir Arthur Pierse in February 1942.

Later, at the Quebec Conference in 1944, it was stated, inter alia, that when weather or tactical situations were unsuitable to attack primary military targets, attacks should be made on industrial areas. Harris knew the limitations of his command in winter months and often resorted to area bombing, in keeping with the general bombing policy.

Let us look at an example of weather conditions as described by Bruce Otton:

We iced up at 16 000 feet. I'd never seen ice before, then, suddenly, I couldn't see out because of ice. The wings were covered with ice and we started going down because of it.

I called up the rear gunner and said, 'You come up here, sit behind me and watch these instruments. Tell me if they start doing anything funny. That is what they what they ought to be doing. Tell me if they're not.' I was circling around trying to find a hole in the clouds so that we could go down. I wasn't that familiar with the area and I didn't want to fly blind because of the mountains. The ice remained until we reached 6000 feet.

Max McVicar had this comment about the weather:
Usually, it was clouded out and visibility was a problem. The country was blacked out and, right from the take off, you were on instruments. The runway lights were hooded and, most of the time, there was no horizon.

There was a story going round of a Sunderland running into a cu-nim in the Bay of Biscay. The upcurrents were so severe that it did a complete loop. The aircraft was severely stressed but they got back.

Fred Wright comments on the changeable nature of the weather:
I remember a trip we did to Worms in February 1945. The target was a synthetic petrol plant.

We took off and flew in cloud for three parts of the way to the target. The target was clear and we could see the smoke rising.

When we left the target we came back into the cloud but that had cleared by the time we got back to base. The weather pattern changed all the way.

Peter Balderston had an interesting experience in fog:
The fog had come in, but we thought we knew where we were, so we called up for permission to land. We were given a time but then found we couldn't see the runway.

They said they'd send up a white Verey light from the end of the runway. We saw the white Vereys, came in and landed. The control tower was calling, 'Where are you Y Yorker?'. We replied, 'We're down. We're going along the runway. We can see a chap ahead of us with a blue light and we are going to follow him.' 'But where are you Y Yorker?' 'We're going into a parking bay now.'

We weren't on our own station. We were on the Beaufighter Unit next door.

Squadron Routine

While details may vary, day-to-day life on a squadron followed much the same routine as that described by Alby Silverstone:
Life on a squadron was quite different to anything we had experienced anywhere else in the Air Force.

Each flying category, i.e., pilot, navigator, etc., had its own section office. Each crew member was required to report to his section by 0900 hours — or 1400 hours if he had flown during the previous night. In the latter case the morning hours were spent in sleeping.

The pilots had a flight commander, usually a squadron leader. The other sections had a specialist officer in their category, generally a flight lieutenant, in charge of their respective groups, i.e. bombing leader, navigation leader etc. The leader was often on his second tour of operations and a member of the senior crew on the squadron.

All signals concerning operations and crews were promulgated in the flight office (the pilots' group). The pilot was, therefore, generally the first to know if his crew was listed for ops and generally contacted the rest of his crew about briefing.

Group Headquarters advised each squadron daily, by coded or scrambled teleprint signal, whether it was required that day for operations or whether it was stood down. In the case of operations the signal would advise the effort (number of aircraft) required and bomb and fuel loads to be carried so that armourers and refuellers could start preparing the aircraft. Target and bombing times were also given.

If operations were scheduled the station went into a complete blackout. No one was allowed on or off the station and telephone lines were switched off.

If there were no operations all aircrew were stood down until reporting time next day and were free to do what they wished.

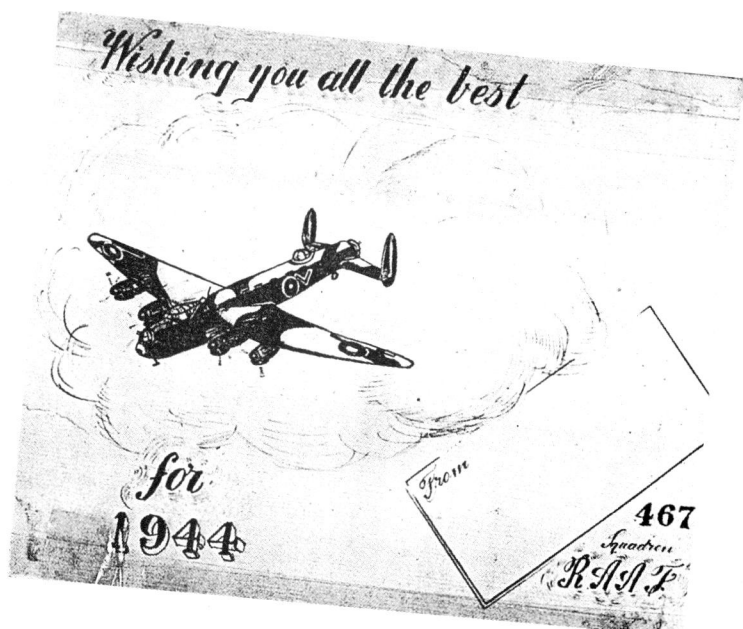

Christmas card, 467 Lancaster Squadron RAAF.

Waddington (467 Squadron RAAF) covered in snow — control tower snowbound.

Operational Fatigue

Sir Arthur Harris, in his introduction to Guy Gibson's autobiography, said: 'Remember that these crews, shining youth on the threshold of life, lived under circumstances of intolerable strain.'

This was undoubtedly the case, and the result showed itself in a number of ways. Ross Pearson, who served on 102 Squadron, recalls that he had commenced to study, as an external student, with the University of London. He gave this up after his fifth trip:

I felt there was little point in studying for the future. The end of a tour was such a far off eventuality. Even if one survived one tour — what then?

I never felt I would not survive — it was always 'the other fellow' who would 'go for a Burton' — yet there were always moments of apprehension when one saw and heard the flak or saw an aircraft going down in flames with no chutes coming out.

At the command level, Ted Eagleton felt the strain:

I was a flight commander — in charge of a flight — and acted as wing commander when he was on a course or on leave. I found the strain quite severe because, in having to choose crews to do a raid, you were really dealing with men's lives. I used to try to give new crews an easy target and use the more experienced on the harder ones.

There were times when, after I had been briefed for a target, the sortie would be cancelled. The next day it might be an easier target. If I was acting wingco I would take myself off and put another crew on to give them two or three easy trips.

But this ended up causing some friction. My navigator said to me, 'Listen Ted! If you ever do that again you can get another navigator.' The other flight commander said to me 'We were both on the original list. You should have gone. It's not fair to your crew.'

Sam Weller had an experience with a 'non-starter':

At times there seemed to be people on a squadron who were unknown quantities. No one seemed to know who they were or what they did. They didn't seem to operate but no one knew whether they were sick and they seemed to stay on.

Our squadron commander decided to take one chap, who had been hanging about the squadron for some time, as a spare bod. The wingco had a fine operational record and was very proud of it. Unfortunately, the wingco came back early on the night he took this chap. He was most upset and bad tempered about the whole thing — the new crew member had said he was ill and so the wingco had come back early.

The wingco said, 'I've never come back early in all my career in the Air Force. This is the first time.' So I said, 'What can you do?' He said, 'I know what I'm going to do. You're going to take him next time.' I said, 'Why should I take him?' He said, 'That's an order. You take him.'

Next night we were on a major German target and I had this new crew member. We crossed the coast and went on towards Switzerland. Suddenly, this chap said, 'You've got to turn round and go back.' I said, 'What's the reason for that?' He said, 'I don't feel well.' I said, 'I'm sorry. I've got news for you. I'm not turning round and I'm not turning back.' He said, 'You'll have to.' 'Oh no!' 'I'm not well, I can't make it.' I said, 'You can go back there and lie on the floor if you like, or, you can bale out. Please yourself.' He suddenly became quite

The rear turret of a Halifax bomber. Ted Eagleton (on right) and rear gunner Barney McCosker, of 466 Squadron RAAF, pose at Elvington, Yorkshire.

well, got up, filled in his log and that was that.

We got back and he reported me to the MO for being unfeeling and various other things. But the MO said, 'I can't help that. This is a case of lacking moral fibre. I'm certain of that.'

But he was a very cunning chap and could pull the wool over people's eyes. In the end he got away with it.

I remember another. He was an Australian in my flight. He was assigned an aircraft but didn't go on the first operation, claiming there was something wrong with the aircraft. He did the same on the next operation. On the third he came back early, again claiming some fault in the aircraft.

By now the ground crew were incensed. They were proud of their aircraft — and their aircrews — and were grossly insulted at the implied slur on their workmanship. The ground crew approached me. They said there was nothing wrong with the aircraft, but there was something wrong with the pilot. What was I going to do about it? I pleaded with them to give him another chance. They did but he did the same thing again.

I said he should be removed from the squadron. It was referred to Group Headquarters but the group commander, a New Zealander, said, 'No Australian will be removed for being LMF.'

He had to be kept on — more or less saying something was wrong with us. When an opportunity came to transfer him we took it.

Jim Tudberry recalls that there was an occasion on his squadron when a pilot 'froze' on the controls while over the target.

Harry Brabin recalls a case which seems quite unfair:
The crew who shared our hut were shot down as we were flying next to one another. The fighter then turned on us but Sandy was able to return some very effective fire. All the other crew were lost except for the rear gunner. He found himself in the sea, about fifty miles from the English coast in the dark of the night.

A destroyer caught sight of him parachuting down and had him safely on board in twenty minutes. He was lucky to survive so long in the freezing water.

The following day, the wing commander rang the sergeants' mess and told him he must fly again that night as a replacement with a strange crew.

Still in a state of shock, his impolite refusal was not well accepted by the wing commander. He was charged with Lack of Moral Fibre, stripped of his rank and sent to an Air Force prison.

Empty bunks were a constant reminder of the 'chop rate'. They did not augur well for survival and had a chilling effect on aircrew nerves.

Alby Silverstone gives an example:
My first shock when we first joined the squadron was to find that my original crew at OTU were there and had completed sixteen operations. They were now half-way through a tour and I was only just starting out.

My second was our introduction to the possible results of operational flying on the first night there. It was very chilling.

We were assigned to quarters in a large brick, dormitory-style building with several wings. Our wing housed two crews, twelve men in the main room and two in an annexe. Aub and Jim took the annexe and the rest of us moved into the main room.

An operation was scheduled. The other crew in our wing, who were scheduled to fly, were idly playing cards to while away the time while waiting. We watched them play and chatted. Later, when they left to fly, we wandered round to take in the atmosphere and, after watching the take-off, went to bed.

Early the next morning we were awakened by the Service Police as they collected the clothes and other gear of our room-mates. They had failed to return from the operation. A very chilly introduction.

Harry Brabin:
There was a young Scottish air gunner on the squadron. He looked so fresh and healthy that it seemed as though a light shone from within. He went on leave just before the final trip of his tour and decided to get married then and there.

However, he didn't return, much to our sorrow. It is his face that I always bring to mind when I hear the 'Last Post' played.

RAAF Halifaxes at Leconsfield.

One of the horrors which we didn't think much about was being burned. Occasionally it was brought vividly home to us. Harry Brabin again:

At Australia House in London I saw a pilot who had been badly burnt on the face and hands. I only just recognised him as a fellow who used to travel on the same train to the city as me before the war.

The petrol tank behind the engine on his aircraft had exploded. His flying helmet had protected his head and hair, but his face was covered in scar tissue.

I saw him again many years later, in Sydney. He looked quite distinguished and successful and his facial skin was quite smooth.

The Key to Survival

The number of crews who finished a tour of operations (usually thirty operations) was not high. Those who did survive were the lucky ones. Some crews were highly experienced yet did not finish. Some were bad and undisciplined and finished. However, the chances of survival were higher if the crew was well disciplined and each member did his job thoroughly.

Crews which survived the first ten trips seemed to have a very good chance of getting through the next ten or fifteen. But the last few seemed to be particularly dangerous.

The length of time crews remained on operational units varied from time to time.

In Bomber Command a tour was ultimately set at about thirty operations (thirty-six in 5 Group). A second tour was twenty operations and a third tour was ten.

Later in the war, as the Germans retreated, a points system

was introduced. A first tour required 120 points. Four points were allocated for a German target and three for a French. This system did not apply in 5 Group.

Given that losses were high, how did crews survive? What was the recipe? Two suggestions have been put to me — one by Sam Weller and one by Bruce Otton.

Sam Weller successfully completed two tours:

A tremendous amount of luck was involved. I had friends who were better pilots than I but they were shot down.

I think an amount of cunning helped. I had one or two ploys. I never came straight up to the coast. When you were coming back across France you could see the pundits [light beacons in England] flashing and you thought you were home you could easily lower your guard. The fighters used to shoot down a lot at this point.

When I came to the coast I always turned 90 deg. (if I was steering 360 deg. I turned to 270 deg.) and then would suddenly whip the nose down and dive across the coast.

I didn't weave [jink from side to side]. My theory was that every time you weaved you lost time — in effect, you flew more slowly.

Bruce Otton:

Luck played a big part in survival but crew discipline and common sense also played a big part.

We survived because we had a top line bomb aimer and a very sound navigator who was like a father to the crew — particularly to the gunners. They were very young but were keen about their jobs and very good at fighter affiliation.

As for me, I was not prepared to allow idle chatter during flight. I wanted everyone on maximum alert.

Where the route was over some heavily fortified place I went round it if I could. I can well remember our squadron commander led a gaggle straight over one such place and was badly shot up. He lost his hydraulics and had to have someone fly beside him all the way home.

There was no need to stick your neck out unnecessarily in my view.

Humour

There were some occasions which, while serious at the time, now have a humorous side. It is surprising how some people can see a lighter side in even the most serious situation.

Jack Davenport recalls one incident:

A navigator friend of mine was very religious. On one trip they were caught in searchlights and having a pretty rough time. They were carrying out violent evasive action when one of the crew called him up and said, 'Are you praying Kevin?' He said, 'Yes. I'm praying.'

Sam Weller as a Wing Commander with 158 Squadron RAF, Lissett, Yorkshire.

A little while later the gunner called and asked, 'Are you still praying Kevin?' 'Yes, I'm praying like Hell!' 'Well shut up and give me a go. You're not getting through.'

Jim Tudberry:

In the wireless school they apparently said 'bale out' if they had a problem. In one instance, the wireless operator was off the intercom but came back on just in time to hear the flight engineer and me discussing a problem. He yelled out 'bale out.' I immediately jumped down his throat, just in time to stop the rear gunner baling out.

THE ROLES OF THE CREW MEMBERS

Before I reach the conclusion of this section, I think it is necessary to give an account of the role of each member of the crew, with the exception of the pilot. I have shown how crews operated. Now, as a finale, I'll show the solo efforts of the bomb aimer, navigator, flight Engineer, wireless operator, gunners (mid-upper, rear, mid-under).

The Bomb Aimer

In 1942, bomb aiming was recognised as a specialist category and the 'B' wing started to become a regular symbol in aircrew.

Bomb aimers, besides having a key role over the target, were also responsible for other tasks. Map-reading and assistance with Gee and H2S were part of their general duties. In some aircraft they acted as second pilots.

Fred Wright, a pilot of an operational Halifax elaborates:

We bombed on the T.I.'s (Target Indicators laid by the Pathfinder Force) all the time and got good aiming point photographs especially due to the bomb aimer. [This is a European theatre view. There are many areas of operations where path finding was non-existent and the bomb aimer was vital in identifying the aiming point — and the target.] He played a very vital part, if only for a few minutes. [Again a European view, largely due to the fact that Bomber Command operated mainly at night. The bomb aimer could not carry out the very useful task of map reading and getting visual fixes that he could in other theatres.]

He was our second pilot as well. He could fly the plane just well enough to get back to England, if necessary, and, over the target, the plane was effectively under his command for the fine course corrections required to make an accurate bombing run.

Bombing up a Lancaster.

The Navigator

The navigator was responsible for guiding the aircraft to the target and home again. The traditional method of navigation involved finding the aircraft's position by visual sighting of the ground and use of a map, supplemented by radio bearings and astro-navigation.

These methods were still effective in many areas but the concentration of Bomber Command's effort on night operations soon showed them to be deficient for those operations. The whole of Europe was completely blacked out at night, making map reading effectively impossible, except in a few special situations, and the weather was seldom good enough to allow regular astro observations. With most fixed radio transmissions blacked out or doctored to prevent their use for radio direction finding, the normal methods were ineffective — at least at night.

The picture slowly changed with the development and introduction of radar type aids such as Gee and H2S. These were a great help but were subject to countermeasures by the enemy. Gee was jammed and its use largely restricted to England until after the Normandy landings, when it expanded

to the eastern border of France. H2S had to be withdrawn from service when the German night fighters began homing on it to attack the bombers using it. Loran was introduced in the closing months. It had not been jammed by the end of the war, but was not as easy to use as Gee.

So navigation in Europe remained, as it was in other theatres, to a large extent an art which required great application and concentration, together with some intuition, on the part of the navigator.

Noel Gilmour explains some of the navigator's art:
Single seat aircraft were usually restricted to a local area and the pilot navigated by map-reading. In Bomber Command we did it rather more scientifically. The navigator found what the winds were and then told the pilot the course to steer to reach the objective.

He plotted on a chart where the aircraft would be if there were no wind. This could be calculated from the course flown, the airspeed and the time. By 1944, most Bomber Command aircraft had air position indicators which did this automatically. A starting position was dialled in and it ticked up the latitude and longitude as you flew — on the assumption of no wind.

If you could get a fix, you could plot the difference between the two positions and calculate the speed and direction of the wind which caused the difference. This was used to correct the course flown as required. With an aid like Gee this could be done every six minutes.

Gee was only available as far as the Ruhr. Normally it could only be used as far as the enemy coast; then the picture became quite unreadable due to jamming.

Radio was not much use because there were no useable radio transmissions. The radio operator could call up some stations and get fixes from them, but that meant breaking radio silence — you had to be in real trouble to use that.

You hoped that, by the time you had lost your Gee, assuming you didn't have H2S, you had a reasonable plot and had a couple of good winds to give some idea of the accuracy of the predicted winds. From there on it was dead reckoning.

I don't remember any time when two targets were so close to one another that, in normal circumstances, you would go to the wrong one. In any case, from August 1942, the Pathfinders would be there and mark the target.

One of the things about being a navigator was that you were working like mad all the time, even when you got out of Gee range. You were always thinking of how to locate your position, calculating winds, updating your plot. You really had your work cut out just after the target.

After you had gone through the target and bombed, you usually had two to four short legs at different speeds. Sometimes you were diving to a different level to put the fighters off, or there might be 90-degree changes of course for the first ten or fifteen minutes after the target. It could be exciting at that time and accurate flying was not as important as safety. From the navigator's point of view. though, accuracy of navigation depends on accurate flying.

At these times the air speed indicator fitted in the navigator's compartment was a great help. It enabled you to keep a check on the speeds flown so that you could keep your plot up to date and accurate. We still maintained our own plot in spite of having the air position indicator.

I operated our H2S. The bomb aimer was supposed to do this but ours didn't take to it. I don't think he'd been properly trained on it. Quite often, however, there were no specific landmarks to be seen. The screen just looked a heap of mush and you couldn't tell where you were. If there were a couple of reasonable towns it was a help. You could get an accurate position and restart your airplot from that.

We took H2S on a test flight. We did a cross-country up north, through Scotland, down the west coast, crossed back in about Fleetwood in Lancashire and back to Driffield — a fairly long trip of about four and a half hours.

We had an Air Taining Corps cadet on board for a ride. He came from Liverpool and had the broadest and most irritating Lancashire accent. He was about fifteen.

We took off from Driffield in a brand new aircraft, climbed to 8000 feet and chuffed off towards Scotland. All of a sudden, the aircraft became full of smoke. The youngster was in a state of panic — he thought the aircraft was on fire. We knew where it was coming from but didn't know the cause. It was coming from the heating system. After some nervous moments we decided it was probably the new paint in the heating ducts.

The young chap kept asking where we were. I, being an obliging character, would tell him. But a couple of minutes later he would ask again. This went on and on and on for some time and I got fed up with it.

Coming back from the Hebrides I said to him, 'Why don't you go down and see Brian, our rear gunner, in the rear turret? Why don't you sit in the turret?'

Brian was quite happy about this. Time went by, and more time went by, and I thought 'Where's that little bloke?' I called him up on the intercom but he didn't answer. I asked Brian if he'd seen him, but he hadn't.

We were now at 14 000 feet and we found him in the rest station, half flaked out. He hadn't got any further because he hadn't taken an oxygen bottle with him. We put him on one and he came round.

We thought it rather funny, but it could have been serious.

The Flight Engineer

With the introduction of the four-engined heavies, Bomber Command started training flight engineers. They joined their crews at Heavy Conversion Units, where the crews first made the acquaintance of the four-engined machines.

The advent of the four-engined bomber brought a number of extra gauges to be watched and a number of new systems to be monitored. Not the least of these was the fuel system.

Some aircraft adopted a system involving a large number of small tanks — the philosophy was that, if one was holed, the maximum loss would be limited to the size of that tank. If one was on the ball, this could be reduced by running all engines from that tank until it emptied and then switching to undamaged tanks. This led to complicated systems with many cross-connections and required careful management to ensure that you didn't try to run the engines from an empty tank. *It also needed fuel to be moved from tank to tank because not all could be directly connected to the engines. This became a nearly full-time job, as good records had to be kept so that you knew where the fuel was.*

It was too much for a pilot to cope with on his own and so the flight engineer took responsibility for it. In addition, he monitored the engine gauges and often assisted the pilot at critical times such as take-off and landing. He had a small kit of tools and could carry out some minor repairs if required. In his spare time he was a valuable extra pair of eyes.

Max McVicar gives an insight into the duties of the flight engineer.

I picked up my flight engineer at Heavy Conversion Unit. He had not flown very much. In fact, he was only eighteen.

When we first went to the squadron we did a cross country on an early model Halifax — the old square wing tip job. It didn't have an effective oil cooling system; if you struck extremely cold weather the oil would coagulate and wouldn't flow properly, resulting in engine failure.

It was snowing on this occasion and the outside temperature was very low. On the Halifax you adjusted airspeed by the engine revs. The throttle was set for cruising and the revs brought down to maintain the required speed.

Halfway through the trip, the engineer told me the oil temperatures were dropping. The engines were not generating enough heat to maintain the oil temperature. If it was allowed to continue the engine temperatures would suddenly shoot up. The effect was known as 'coreing.'

I had to increase revs to 2500 rpm. This caught it in time and we brought the oil temperature under control. If the engineer hadn't been watching the situation we could have lost all four engines and had a disaster.

The Wireless Operator

The role of the Wireless Operator varied considerably, depending, as with other crew members, on the command in which he served. Two Bomber Command experiences, however, are indicative of the type of work and the conditions of operation.

A pilot's viewpoint is given by Max McVicar:
The addition of H2S made another chore for the wireless operator [Sometimes the bomb aimer operated this].

On operations, there was radio silence from the aircraft but the wireless operator was getting messages from base, even while over enemy territory. He also watched his 'Fishpond'.[5] Any unusual move of aircraft around us was

reported. If they were relatively stable, okay, but if the blip moved suspiciously we took evasive action.

Harry Brabin, a wireless operator, says something of what operations meant to a wireless operator:
In my swivel chair, directly below the pilot, I afforded good protection to Bill. Fighter pilots had steel plates fitted under their seats — Bill had me!

There was a small window near me so I could see out. The Marconi transmitter and receiver were on top of one another in front of me and I had a small folding table fitted with a morse key. The table had to be lifted out of the way when I left my position. It was very cramped.

We did not speak to one another much on the intercom as Bill did not encourage idle chatter. I could tune in to English radio stations for music or news but I seldom did.

Ross Pearson:
Sometimes I wondered why we carried a wireless operator — it seemed as though I was a passenger taken along for the ride. Then I started to take stock of what I did or could do.

First, there was the listening watch. This involved tuning in to either a Bomber Command or a Group frequency, or both. It was vital to pick up any recall signal — a single aircraft over the target would be very lonely, vulnerable and unlikely to survive.

I listened intently, hoping all the time to hear a recall; especially when headed for the more unpleasant targets such as Gelsenkirchen, Hanover or Magdeburg.

Next, it was important to pick up bombing wind to be used by the bomb aimer.

Finally, we listened out to know what was ahead of us on return. Was there fog over our drome? If so, to what drome were we diverted? It was important to get any information broadcast about changes in atmospheric pressure. Base provided this information so that the pilot could adjust his altimeter. A friend of mine was lucky to survive. His wireless operator missed the correction and the plane flew into the ground while descending through cloud.

We transmitted very little. Wireless silence was usually very strictly observed. The exceptions to this were, if you were a wind-finding aircraft, if you were about to ditch or were lost over Britain.

It was possible, by using M/F D/F stations, to get a fix. This was used if you were about to ditch. You sent

---

5. Fishpond — a radar aid very similar to H2S. Radar beams were sent downwards from the aircraft and any reflections were displayed on a cathode ray tube. Whereas H2S was designed to pick up ground echoes, Fishpond was designed to pick up echoes from aircraft. Thus a watch was kept for night fighters sneaking up under the bomber, in its blind spot.

out SOS and clamped your key down. If you were lucky, rescuers could get a fix on your position and send Air Sea Rescue to pick you up. On one occasion, late in 1944, the M/F station at Hull had some thirteen SOS calls in fifteen minutes.

The Gunner

The situation of the gunner varied according to the aircraft in which he flew. He generally occupied a turret in either the tail or the dorsal (mid-upper) positions. Front gunners were not used in RAF aircraft but there were a few mid-under gunners in some Bomber Command squadrons. The gunners in aircraft which carried more than one gunner, rarely, if ever, swapped their positions.

A gunner had much to contend with. There was the sheer loneliness of the turret. He was perched out at the back of the plane or seated above or below it. He was vulnerable to attack, subject to extremes of cold and knew that the lives of his crew depended on his vigilance.

Keith Spain:

I only fired my guns in anger once. It was at a Mosquito which should have been marking the target 10 000 feet below us. It came up to have a look. I saw this aircraft coming at us and gave it a few squirts. I missed.

I could see the flak vividly from the turret. On one raid, over Kiel, the ack-ack seemed to come up through a hose pipe. The tracer was constant. I think they were rockets because the tracer came one after another and I don't think they could load them that fast.

It was 20 deg below freezing but we had electric suits. They weren't always effective. Mine fused a couple of times and burnt a hole in my pants. We had silk underwear and gloves with leather gloves to wear over the top. It was a court-martial offence to get frostbitten so you always wore adequate clothing.

In spite of the cold, we didn't have trouble with the guns. Loading was no trouble, we had a toggle which let us cock the guns easily.

We had 'wakey wakey' pills to keep us awake. They were okay but I remember one trip was cancelled after we were in the aircraft and had taken our pills. I couldn't get to sleep for four or five hours.

Ted Priest:

We flew in extremely low temperatures — as low as forty-three degrees below freezing. We had heated suits but if they conked out you were in big trouble.

Arthur Doubleday:

There was a case, when I was acting wingco at 467, when a gunner was found dead in his turret, still hanging on to the grips, when the crew returned from an op. His oxygen had failed, probably right from the start.

It was unfortunately true that rear gunners were knocked out — they did draw the fire. When I first went on ops there were gory tales about hosing the gunner out of the rear turret but they were pure fable.

This is an interesting observation by Doubleday about one of the myths of Bomber Command. Contrary to the commonly held view, gunners did not suffer the highest casualty rates. Indeed twice as many wireless operators (2328) were recorded killed than were gunners (1210).

6

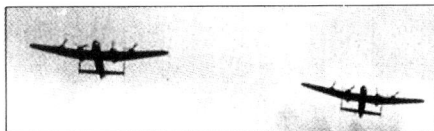

PRISONERS OF WAR

[European Theatre]

In putting together this section of the book I have opted to tell, as far as possible, the continuous chronological story of some Air Force prisoners of war.

My starting point is always just before, or simultaneous with, the actual landing.

Most of the airmen who became prisoners of war came from Bomber Command. This is not surprising, given the number engaged in bombing and the fact that, by and large, they operated over land, whereas those with Coastal Command were over the sea.

The chances of a crew getting safely out of a stricken aircraft were not particularly good for those of either command, but were probably rather less for the Coastal boys as they were generally at lower altitude. Having got out, the chances of being picked up from the sea, particularly in the northern winter, were not good.

Approximately 20 000 members of the RAAF flew with Bomber Command. Of these, some 4000 became casualties and 1150 became prisoners of war. The stories which follow all relate to members of bomber crews who were able to escape from stricken aircraft.

Keith Skidmore was a mid-upper gunner who was shot down in an attack on Rhemshied on 30/31 July 1943. The story of the attack is told in Chapter 4. Here we pursue the story after he landed on German soil:

My first thought was, 'Where am I?' I didn't have a clue. Fortunately, I had landed in a field. Fifty metres to the right and I would have landed in a wood of tall trees.

You were told to bury your parachute and get out of the area should you have to bale out. This was easier said than done. It was impossible with bare hands.

I rolled the parachute up and tried to conceal it. Eventually, I hid it under the trunk of a fallen tree in the woods.

Gee! Was I tired! I lay down in the woods and went to sleep.

On awakening, I got a hell of a shock. I was covered with field mice — hundreds of them I reckon. I jumped up and down for about twenty minutes before I got rid of them. If only I'd known I was going to fall in a river in about ten minutes, I could have waited and drowned them all in a couple of seconds.

What was I going to do? I opened my escape kit and checked it out. Boy! Was I hungry and thirsty? As I wandered through the wood I saw a river and, as I got closer, the trees thinned out. I could see a bridge about 500 metres away. I decided to crawl to the river's edge. To my surprise, the river had a concrete embankment. The water was about a half metre below. I bent over to get a drink and fell in. Gee, I wished I had eaten the biscuits in my escape kit first. I lost it.

I took quite a while to get out of the river and my leg stung quite a bit as I crawled back into the woods. I found out, about a year later, I had a couple of splinters from the aircraft in my leg.

I had to make a plan of how I was going to escape. I decided I'd pretend to be a kid. I weighed only nine stone odd and was of small stature.

I discarded my battledress jacket, cut my flying boots off at the heels and made shorts out of my battledress pants. Gee, did I make a mistake! The grip of the flying boots is on the calf of the leg. I couldn't keep the damn things on. I had to shuffle about. I tore the tail off my shirt, made a couple of strips and wound them round the boots. That didn't work too well.

Eventually, I found a roadway and, as I looked along it, two boys came along on bicycles. If I could get a bicycle it would overcome my shoe problem because both feet would be on the pedals. It was like looking for a needle in a haystack.

Where I was appeared to be a farming area, but I could not see any farmhouses or people, so I just shuffled on. Approximately two hours later I found a road sign with the letters KOLN on it. I sneaked into a wood at the side of the road to examine my escape map. No luck — no such town anywhere. Months later I learnt that KOLN was the German spelling for COLOGNE — my map had Cologne on it.

As I was shuffling out of the woods I found an old scooter near a stone fence. I carried it to the roadway, got on it, and, with the first push, my shoe came off and so did the rubber tyre on the front wheel. I discarded the scooter as it made too much noise without the tyre and I didn't want to draw any attention to myself.

I was quite frustrated and very hungry as I shuffled on. Then, out of the blue, I saw what appeared to be a grocery shop — way out in the country, all by itself. I cased the shop for a while and then decided I'd go inside and get something to eat. We were issued with Belgian money in the escape kit.

When I got inside, I realised that the shop was also a restaurant. On the counter there appeared to be take-away food — cakes and pies.

A woman came to the counter and I pointed to the what I wanted. I made out I couldn't speak as I was burnt around the face and my lips were pretty swollen. I gave her the money and she went to the till. She came back with her husband, who had a revolver. He flashed it at me and I put my hands up. I must have needed ration tickets. I'd never thought of that.

He made me empty my pockets. In the meantime, his wife had left the room. When she came back she had her daughter with her. The daughter was a nurse, had done her initial training in a London hospital, and spoke very good English.

She examined my face and told me not to worry — my burns were not really serious. She excused herself, left, and then came back with a tube of Tanafax. After all this, I knew I was not in Belgium or France. It was Germany. Then the owner of the store took me to the local village (Liedesheim), which was a fair distance away.

On arrival, I was handed over to the mayor. We couldn't understand one another, but he gave me a very good meal. I was put into the local gaol for a couple of days until the Luftwaffe guard came for me.

Then we went off to Frankfurt. We changed trains a few times prior to getting to Frankfurt and Stalag Luft 3. It was on one of those changes that I had the most amazing coincidence I've ever experienced in my life. The guard put me in the station master's office while he went away to get something to eat.

A couple of minutes later a woman came up and put her head against the barred window. 'You're an Australian?' she said. How she knew I'll never know. She must have known the guard or the station master.

She told me she had been in Australia for many years — her husband was an engineer with Dorman Long when they built the Harbour Bridge. My cousin's husband was an English architect with the same Dorman Long on the same project. I told her so and she asked me his name, Walter Goodsmith. She said she had met him.

Bringing out the stew, Luckenwalde POW camp, May 1945.

U.S. tented POW camp, Luckenwalde.

When I saw Walter, after the war, I told him about my experience and he asked me her name. I never thought of asking her name. He told me there were many German engineers on the bridge project, so, without her name he could not identify her. He said they must have met at social evenings.

Seventy-eight of us went by train from Frankfurt, across Germany to the little village of Muhlberg. When we got to the camp we were located in our own compound. There were many other compounds, and each housed a different nationality. We were the first British prisoners to arrive in this camp.

The food was shocking. They used to give us a lump of black bread and ersatz cheese. We didn't get it every day. It would often have to last a couple of days.

At night we got a skilly. It was supposed to be a stew. It wasn't very thick, occasionally you got some meat. On one occasion I got an eye. We didn't know whether it was an ox.

We did get some kind of tea.

When the Red Cross parcels eventually came our food was supplemented. But it took eight weeks for the Germans to notify the Red Cross about us and for the first parcels to arrive. There was three weeks supply for the original seventy eight prisoners in the first batch of parcels. But, by then, the number of British prisoners had increased to one hundred and ten or twenty so the supply was inadequate.

Someone in England, a Rotarian, decided to adopt me because he thought I'd get parcels more quickly from England than from Australia. But I got a lot of stuff from Australia — letters and fruit cake. The chap in England sent me cigarettes, which I used to bribe the guards to get things for me.

There was an army station not far from us which, apparently, had an ammunition dump. One day there was a tremendous explosion. It blew out all the windows in our huts.

We had one bloke with us who always ate by himself. He used to eat his goodies in front of us. On the day of the explosion he had made a big plum pudding out of raisins and biscuits. It was sitting up on the window sill and all the glass went into his pudding. It spoilt his party but we weren't sorry.

To pass the time, as I could draw, I used to scribble a bit on the walls. We could also play darts or quoits if we wished. We had to stop playing football because the compound had, at some stage, some sort of pit which had food buried in it. Some of the officers were worried that it could have been diseased. However, we did play basketball.

One day I was late getting out on the parade ground to be counted. The head German guard, who used to count us, came into the hut and found me still there. As a penalty, I had to do some camp work. Normally, NCO prisoners were not required to work. He came back with a big axe. He said I had to break up some stones and lay them in front of the hut. There was a pool of water there and the stones would save us walking through the water.

I hit the stones with the cutting edge, rather than the back, of the axe. He wasn't very impressed.

Luckenwalde POW camp at the end of the war in Europe, May 1945.

I was late again, about six months later. I was crook — I had pleurisy. Anyhow, I went out and had to do duty — mend the barbed-wire fence. He picked up the hammer and got me some nails. He held the nail and I was to hit it. I hit his hand. He shook his head. I finished the job anyway.

On 26 January 1945 I wasn't late, but he called for volunteers to do some work. You wouldn't believe it. I put my hand up and nobody else did. He just looked at me and shook his head.

Certainly, I had volunteered. So he took me into the hut and let me have a cup of tea and a cigarette while all the others were out in the cold — but he wouldn't trust me with any work. The other blokes called out, 'Collaborator!'

I was very lucky. I was in hospital for eight weeks. I had pain and they took me up to the doctor — I don't remember whether it was the German or our own doctor. They shoved me in the lazarette for eight weeks and I was treated well during that period. That coincided with the Normandy landings. The other prisoners had a very rough trot in that period but I ate well.

There was one bloke who mucked in with me who used to disappear every now and then. He escaped about three or four times. He would always be back after a couple of days.

He told us he had once got on a bus and got down behind a seat. A German officer and his wife got on and, with the whole darn bus to choose from, they walked down to the back of the bus and sat on top of him.

Eventually, a French POW, who was about to be repatriated to Britain, died. This fellow took his place. He could speak French, and was successful.

The Russians liberated us on 23 April 1945. We then had to make our own way home. There were no Jerries in the camp, except for the commandant. They all shot through because the Russians were coming one way and the Americans the other.

The Russians, on their arrival, shot the commandant.

We shot through as soon as the Russians arrived because there was no food. The Russians were battling for food themselves.

A mate and I had to walk. We had to cross the Elbe — the Americans were on one side and the Russians on the other. We found a broken down bridge where we could cross.

Before we were allowed to cross, the Russians took everything off us — including our watches. Once on the other side, the Yanks took us to Halle by truck. It was the first time I had seen any real war damage. They flew us to Brussels, where we stayed about a week, and then flew us back to England.

Bruce Loane was a wireless air gunner with 466 Squadron. His experiences during the fighter attack on his aircraft are described in Chapter 4.

His experiences on and after landing on 21 December 1943 are now described:

It was very cold in Germany at that time of year and I looked around, trying to think what I could do. I had hurt my head and lost the sight of one eye and my vision was blurred. I knew the troops would be out looking for me and was lying on my back, looking left and right, when I saw what I imagined were troops carrying torches and looking for shot down airmen. It was a bit of a problem. What to do next?

Then I found that my Mae West had inflated and the little light on it was moving as I moved. It was that light that I could see and not troops. I ripped it out and took off the Mae West.

I buried the chute and tried to work out what to do. My immediate problem was that I had no boots. I'd come on ops with only one because I had hurt my foot while playing rugby. Somehow, while coming down, I'd lost the other one. I had only a light pair of socks and my feet were beginning to freeze.

We had escape kits, but these were fairly primitive. In the early light there was 9/10s cloud so I couldn't even navigate by the stars. I couldn't read the escape map because of my eye problems and, although I had a compass, it was useless in the dark.

I used to spend a fair amount of time in the squadron Intelligence section and thought I might just fluke recognising something familiar in the area. If I could get on a road there was a chance I could find a signpost to a major city. I was peering about for a sign of a road when the equivalent of the Home Guard, the Volkstrum, came up and suggested I come with them. I got the distinct impression they were not joking.

I was questioned near my place of initial capture by a girl interpreter. She said, 'They are getting very angry with you for dodging the questions.' I said, 'You don't have to tell me that, I can see it. Tell them you can't make me understand.' Then she said, 'They want to know where you buried your parachute. Will you tell them?' I said, 'Yes.' But I had no idea where it was.

We went out to the little village near where I claimed to have buried the chute. On the way she and I were having an argument about who started the war. She called me a Luft gangster.

Finally she gave up because I claimed I couldn't remember where I'd buried the chute. She passed me on to another gaol.

I went through a series of gaols and finally got to Frankfurt proper. A lot of those who baled out didn't make it that far. I believe less than 10 per cent reached prison camp. If not killed in the air, they were killed on the ground — by Germans, military or civil. It was to be expected at that time.

Frankfurt was the main interrogation centre. They were very efficient and very smart. They would put you in a highly heated cell and disorient you so you had no idea where you were. There was no way you could even tap morse.

They would pull you out at all sorts of times. The guards would just come in and haul you off to interrogation. It's pretty hard to just keep saying, 'Under the Geneva Convention I only have to give my name and my number.'

I remember an impeccable German officer at the Interrogation Centre. He was seated, drinking wine and breaking French bread while I was standing there with no boots and feeling fairly unhappy. He asked if I'd like a cigarette. I wasn't going to be gracious so I refused.

He said, 'What is your name?' I was in battle dress but he said, 'Where are your identity discs?' (He knew I didn't have them.) 'Well,' he said, (His English was better than mine — I believe he was educated at Edinburgh University) 'If you are an officer, as you claim, you would know you should have your identity discs at all times. You could be a spy, which you probably are.' I looked less like a spy than anyone you ever saw.

Suddenly, he said, 'How was the Gee? How was the H2S?'

I tried not to be surprised, but I nearly fell over when he talked of the H2S. I think we were one of the early squadrons with it. It was still in the experimental stages, I believe. In the room were all the things that shouldn't have been there; documents, and even a book of instructions that could only be read by officers in the company of another officer. I had only read it a few weeks earlier and it would not have been available to everybody.

We'd always been instructed not to take restricted information into the air. Obviously, not everyone had obeyed the orders.

I don't know how long I was at Frankfurt, but I was eventually sent to Stalag Luft III. I was put into a compound called Belaria. It was the latest one established.

Stalag Luft III was where the Great Escape was carried out.

I kept a log book and in it is a memorial service to all those who escaped and were subsequently shot. All the Germans were standing round us at the service. They were a bit uneasy, armed of course. I remember the Senior British Officer saying, in front of all these German guards, 'Remember these bastards! Any of you that live must never forget.'

The escape organisation was very good but needed hundreds and hundreds of helpers. There were camp guards in all sorts of positions, as well as roving guards moving about. The Germans expected us to try to escape; we had said we would if we got the opportunity.

We asked the Germans to allow us to have German newspapers. They thought it a good idea because they

THE LOG

10th April, 1944

BELARIA

In Memoriam

THE CAMP WILL HAVE BEEN SHOCKED TO HEAR OF THE DEATH OF *FORTY-ONE OF OUR COMRADES WHO DIED RECENTLY IN THE PERFORMANCE OF THEIR DUTY. AS A MARK OF OUR ESTEEM, AFFECTION AND RESPECT, IT HAS BEEN DECIDED THAT ALL ENTERTAINMENTS WILL CEASE UNTIL A MEMORIAL SERVICE HAS BEEN HELD.

THIS SERVICE WILL TAKE THE FORM OF A PARADE SERVICE FOR THE ENTIRE CAMP AND WILL BE HELD ON THE SPORTS GROUND IMMEDIATELY AFTER MORNING APPELL ON THURSDAY, 13TH APRIL ; OR, IF WET, AFTER THE FIRST FINE APPELL SUBSEQUENTLY.

" They shall grow not old, as we that are left grow old :

Age shall not weary them, nor the years condemn. At the going down of the sun and in the morning We will remember them."

SENIOR BRITISH OFFICER

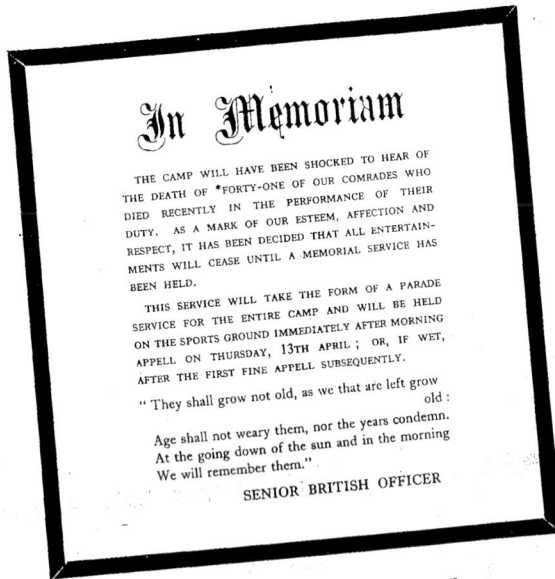

* This tragic figure was subsequently learnt to be fifty-one.—ED.

(4)

Memorial notice for Great Escape in *The Log*, POW paper circulating in Belaria camp.

NOTICE

INFORMATION HAS BEEN RECEIVED FROM THE GERMAN AUTHORITIES THAT CERTAIN AREAS IN GERMANY, APPARENTLY IN THE VICINITY OF THEIR ARMAMENT AND WAR INDUSTRIES, HAVE BEEN PRESCRIBED IN WHICH ANY UNAUTHORISED PERSON IS LIABLE TO BE SHOT ON SIGHT.

A REQUEST FOR MAPS SHOWING THESE AREAS HAS BEEN REFUSED ON THE GROUNDS THAT IT IS UNDESIRABLE FOR THE GERMAN AUTHORITIES TO PUBLISH THE AREAS IN WHICH THEIR ARMAMENT INDUSTRIES ARE SITUATED.

ANY PRISONER OF WAR WHO ESCAPES OR CONTEMPLATES AN ESCAPE IS ADVISED TO AVOID SUCH AREAS IF POSSIBLE.

28th September, 1944.

GROUP CAPTAIN, Senior British Officer.

This notice was publicly displayed, much to the bewilderment of the German Staff who took it all very seriously—ED.

(101)

Ironic notice posted in Stalag Luft III POW camp. Belaria

were full of their propaganda. We had our own lookouts. When we got the nod from our lookouts that the commandant was coming, we would have the papers open in front of us. The flunkey used to come in and was impressed with the navigator who was teaching us how to pronounce German words: 'When you are pronouncing this word, you have the cheeks of your arse like this.'

If they had turned over the pages they would have seen we were copying maps for escape purposes. We did ask for maps — without success.

Digging tunnels occupied a number of people because you had to get rid of the dirt. The guards were on the lookout for it all the time. We put the dirt inside our battle dress pants and socks and then put on a bit of a kerfuffle. We would march around in a crazy fashion dropping it.

We were digging a secondary tunnel. The Germans could let it get close to the wire and then break it in — but we would still have one still going.

I was way down the list, but I admired the tenacity of some of the people I was with.

In the beginning we used to get Red Cross parcels. The German food was never very good — they only had a little food themselves. Towards the end, we were served only two slices of black bread per day with some soup — kiwi soup we called it. I have no idea what was in it — rumour had it that you were sent home if you found a potato.

It was pretty grim actually. We had a couple of doctors in the camp and I have recorded in my log book that they said that, if you laid down all day, the food wouldn't be sufficient to keep you going. So those who had been in camp a long time were beginning to feel the effects — even though, as aircrew, they had been pretty fit to start with.

We had limited space, but, as we weren't let outside, we had to exercise around the camp. In the early days we had all sorts of activities.

Ralph Parsons was a bomb aimer in Bruce Loane's crew. He was shot down on 20/21 December 1943. His experience differs from Bruce Loane's, but let him tell the tale:

I landed on the side of a small hill. My chute got tangled

in a fence just above me. I saw the figure of a man carrying a lantern, coming out of a house, which was just discernible, only a few hundred yards away. I lay quite still on the ground until he went back inside.

It was an awkward job to untangle the chute, without making a noise, and while clad in heavy flying gear, on cold, slippery ground. Once it was done, I made as small a bundle of the chute as I could, though it was still bulky and cumbersome, and headed away from the house. It was not long before I literally stumbled upon a mound of earth. It was about three feet high and twelve feet long. I scraped a hole in it and buried the chute. I later learnt that, on the Continent, potatoes were often buried in such mounds for storage during the winter. Some Jerry was in for a surprise when he came to get some spuds.

My next move was to find somewhere to hide and to get my bearings. While trudging across a muddy field, I saw what looked like two figures walking towards me. I flopped down into an icy furrow, hardly daring to raise my head to see if they were still coming.

It was bitterly cold and I took a furtive glance, but they were still there. I was so cramped and cold that I just had to raise my head and have another look. It was then that I discovered they were tree stumps. I think that, perhaps, I was a bit jittery. I chipped the ice off my trousers and made for the nearest clump of trees.

It was an eerie sensation, being alone in a strange and hostile country. It was not hard to have the imagination take over — so much so that I was soon surrounded by wolves, bears and Germans with fierce dogs. Actually, the most dangerous animal I saw was a hare that scampered past me next morning.

I would dearly have loved to have lit a fire to thaw out, but I dared not take the risk of being seen. I found a fallen tree and spent the rest of the night in a depression at its foot.

Before dawn, I took the compass and miniature box of matches from my escape kit. I considered it safe enough to light a match and find my bearings. Having done this, I decided to head east for a while, reckoning that this might give me a better chance of eluding any search parties.

I shed my heavy and conspicuous flying gear. My clothes now consisted of the usual winter underwear, thick woollen socks that came right up to the thighs, a heavy roll-necked pullover, blue battledress jacket and trousers and fleecy-lined flying boots.

I left my pine tree nest and headed east for a while. As it was now getting lighter, I started to look for somewhere to hide during the day. I eventually found a small shed near the edge of a forest. The door opened quite easily, but I had to climb over some boxes to get into the interior of the shed.

When daylight came, I discovered that the boxes were really beehives — probably being stored for the winter. Not being an apiarist, I had no idea how to get some honey without being stung. Some of the hives had glass doors and I could see through them that the bees seemed rather lethargic from the cold. Then I had the bright idea of making them even colder.

I could see that the coast was clear outside, so I sneaked out and gathered some ice off the nearby pools. I gingerly opened one of the glass doors and slid the ice between the frames. After waiting some time, I opened the door again and tried to pull out some of the frames — only to have the bees come out, buzzing around my head. I tried a couple more hives with the same result. I didn't get any honey.

I could see out without being seen, and it was an unreal feeling, able to watch the Germans going about their normal lives, being watched by me, the enemy, and unaware of my presence.

I spent the day studying my escape maps, and occasionally nibbling on the malted milk tablets from the kit. I decided to make a circuit north then west, hoping to get to Belgium or France. As evening approached I made my way towards a village that I had spotted from the shed. I wanted to identify it and reckoned that, in my blue battledress, I wouldn't be too conspicuous.

On nearing the village, I noticed some activity in the railway yards. In the dusk I could just make out several uniformed men. They were searching through the trucks and even prodding under rolling stock with long poles. They could have been looking for me or others hiding or trying to hitch a ride. I soon made myself scarce. I still had no idea of the name of the village. I skirted round it, found a road and headed north.

There was not a lot of traffic on the road, but it was a bit dicey trying to keep out of sight each time a vehicle came along. Most of the traffic seemed to be military. I therefore left the road and headed for where I reckoned the railway was. I found it and made my way north, knowing full well that, if I was caught, I ran the risk of being treated as a saboteur or some such.

Flying boots were not really designed for walking and walking along railway tracks wasn't very easy. I decided that the railway was the Marburg-Kassel line. It suited me now to leave the railway and head westward, keeping, as much as possible, to more open country where I had more chance of finding forests and streams etc., where I might hide during the daytime. This was not all that easy to do. I had to do a fair bit of groping about in the dark and I had to use my compass quite a bit.

After my first night, I found it necessary to carry a long stick to prod the surfaces of pools, etc. to see if the ice would bear my weight.

On one occasion, I came across a stream I had to cross. Unfortunately, there was a sentry box and a striped boom across the road. It was, evidently, the entrance to a prohibited area. I discreetly retraced my steps.

However, I was gaining more confidence and at the nearby village I walked right in and mixed with the few people who were about. I was surprised how easy it was to do. One thing that worried me was that I was hatless and this might be conspicuous in the freezing weather.

I stood alongside a hoarding with posters pasted on it, wondering how I might cross the bridge I wanted to use. The few people about seemed only interested in getting home out of the cold. It was an odd feeling, being almost able to reach out and touch Germans, and I felt almost boastful that they were unaware of my presence.

I decided I was being a bit too cautious about crossing the bridge and set off towards it. I had not gone far when I came face to face with a German soldier. He was a big fellow and wore what looked like an officer's uniform. As sometimes happens in such situations, each stepped to one side to pass, only to find that the other has stepped the wrong way. With an almost automatic reaction, I kind of pushed out my free hand, touched him on the elbow, uttered some odd remark and walked past him.

He did turn round as I passed, possibly thinking that this bare-headed little oddbod, with an oversized walking stick and carrying a dilly bag, was some sort of unfortunate nitwit. The dilly bag was really a satchel from my escape kit — it now contained water. I 'hurried slowly' to get out of his sight and crossed the bridge without any difficulty.

My heels were getting sore and I was wishing I had some sort of transport. My rate of progress was less than I liked. I thought of getting hold of a pushbike somehow. In one town I did spot a bike in front of a house. I sneaked in, only to find that it was chained to a post. I also noticed that it had a numberplate on it. However, my foray was not in vain. I was able to pick some vegetable tops, carrots and beet. I stuffed them inside my battledress. They proved a useful supplement to my rations.

The wailing of air-raid sirens seemed to bolster my spirits a bit. I thought of the crews up there, carrying out their duties, and thought that, in a few hours, they would be back on their squadrons. By the sounds of the raids, I tried to estimate which city was the target — Dortmund, Kassel, Cologne, etc.

Near one village I found a shed where I hoped I might hide and get some sleep during the day. The door was not locked, but the shed was chock-a-block with nets and poles. I had to spend the day leaning against some nets, and only getting short naps off and on. I spotted some paper on the floor, under some poles. It had a lot of red and blue lines on it. I managed to get it out without tearing it, but could not make any sense of it. I thought it might be a map of some sort, but finally worked out that it was a dress-making pattern.

While in this shed, I tore a strip off my shirt and made a bandage for my heels. It was not satisfactory as it creased up and hurt my heels more than the flying boots on their own.

Still looking for some form of transport, I made my way to what looked like an important railway siding. I hid behind the stanchion of a water tower and again saw men searching rolling stock and prodding under trucks. I almost felt like singing out, 'You're looking in the wrong place.'

It was here that I again heard the wailing of air-raid sirens and the sound of an approaching bomber stream. I was hoping I wasn't in the target area. The bombers continued on their way.

My main aim during the day was to see without being seen. In the woods I devised a way to do this. I made two heaps of branches, lay on one and pulled the other on top of me. It didn't work too well. The sticks soon stuck into my back and hips. When I moved to get more comfortable I would spill the top layer off anyway.

Unlike Australia, where the towns and villages may be a hundred miles apart, in Germany they are often within walking distance of one another. They also made much use of cattle as beasts of burden. It was not uncommon to see a farm wagon being pulled by a pair of bullocks, or a horse and a bullock or even a horse and a cow. It therefore came as no surprise when, one night, I saw two cream cans on a wooden platform by the side of the road. I dared not hope that they contained real cream — but they did.

The lids were held on by strong clips. As I undid one clip, it sprang free and made a noise which made a dog bark in a nearby house. I barked back at him while I filled myself and my satchel with cream.

I went on my way but, after going a few hundred yards — and some protests from my stomach, I reckoned I could hold some more cream. So I went back and gorged myself again.

My Christmas dinner for 1943 was limp vegetable tops with sickly malted milk tablets and strong chocolate cubes for dessert.

An eyewitness report of the crash of Ralph Parsons' plane:

Crash of herementioned plane near Beltershausen Landkreis Marburg on the night of 20/21 December 1943.

The approaching fireball was watched by many villagers who had been warned of the air attack on Frankfurt or Giessen.

Approximately over the township of Beltershausen a wing of the aircraft broke off and landed about 80 metres from a big farm in the open field. Because of the fuel it contained an extremely big fire broke out. As the aunt of the owner of this land adjoining the farmhouse saw this great fire, she must have believed that the whole farm would burn up. This shock or fright caused the 50-year-old Margarethe Steiss to suffer a heart attack. Without regaining consciousness this woman died some weeks afterwards.

The aircraft with the pilot crashed about 400 metres behind the village of Beltershausen in an open field.

Parts of the aircraft were scattered in a circle of roughly 1000 metres.

It appeared that the pilot was first buried here, but was later exhumed.

It was apparent that the pilot did everything to avoid the village of Beltershausen.

No further details can be given unfortunately

Bektershausen 17th January 1975.

[According to Parsons; 'The plane was completely uncontrollable from the moment we were hit.']

Parsons was ultimately taken prisoner by a German civilian armed with a rifle. He was then handed over to the military, who transferred him to the interrogation centre at Oberusel. We pick up the story as he is about to leave the centre:

Next morning, after eating my ersatz bread and paste, I was led out to join about fifty other dirty, unshaven inmates. I looked along the line to see if any other crew members were present. My service number was 409226. You can imagine my surprise when I spotted Ian Robinson, 409227, with whom I had done some training in Australia.

We were addressed by a German officer, who spoke with a strong Brooklyn accent. He told us we would be going to a transit camp near Frankfurt. After seeing the bomb damage in Frankfurt, I think most of us were secretly hoping it wouldn't be too close to Frankfurt.

The Yank/German took a photograph of us. We were a sorry, bedraggled looking bunch — a beaut subject for propaganda purposes. He advised us that we would be travelling by tram, there being a shortage of guards, and told us not to try anything silly — like trying to jump off the tram — the guards were good shots and we would be dead before we hit the ground.

At the transit camp we were again searched, fingerprinted and photographed. Then our first contact with the Red Cross.

Once we were installed in the transit camp we were

Beds at Luckenwalde POW camp.

able, thanks to the Red Cross, to wash, shave, and eat some real food. A British officer had been retained by the Germans for liaison purposes. He regulated the supply of food from the Red Cross parcels, according to the number of transient prisoners.

Since practically every prisoner had been through some kind of traumatic experience before capture. it was only natural that being with others of your own kind, just being able to talk freely, gave a great feeling of relief. However, it would have been easy for the Germans to plant infiltrators among us so we tried to be very cautious in our conversations and avoid reference to anything that might be of use to the Germans.

After a few days in the transit camp, about 200 of us were told we would be going to Stalag IVB Muhlberg — situated between Leipzig and Dresden.

I tried to kid myself that the train ride would be a welcome change. After walking, under a heavy guard, to Frankfurt, and seeing the terrible devastation, we arrived at the rail centre. When we saw the overturned engines and trucks I think most of us thought, 'Get us out of here — but quick!'

Our transport was three ancient cattle trucks of French origin. On the side of each truck was the sign '8 chevaux, 40 hommes'. It was pretty obvious that our train journey to Muhlberg was not going to be very cosy. About seventy men, not forty, were crammed into each truck: vertical sardines was an apt description. A couple of bundles of hay were tossed in. This was meant for bedding — provided you could find room to lie down. There was also a toilet facility, a large bucket, in the middle of the truck. Anyone unfortunate enough to be near the bucket was likely to be splashed by its contents.

Each of us had been issued with a Red Cross parcel. These contained such things as Spam, bully beef, powdered milk etc. — in short, the essentials. As there was very little water available, to consume the powdered milk one had to hold a little powder in the mouth until it was moist enough to swallow.

Occasionally, when the train stopped, usually at stations, the guards would let about half a dozen of us jump down on to the tracks to answer a call of nature. Many of us were rotten with dysentery so this was quite spectacular. It usually happened in full view of commuters and modesty was impossible. The guards made quite sure they kept us covered with their tommy guns all the time.

One guard spoke quite good English. He told us that he had a furniture business before the war and traded with British firms. He was an intelligent person and it came as quite a surprise when he referred to 'our Fuhrer' as though he was the ultimate answer to all Germany's problems. But the tone of his voice did not carry much conviction.

After a three-day journey, which would normally have taken one day, it was a relief to climb down from our smelly, cramped, cattle trucks into the fresh air of Neuburxdorf, near Muhlberg. The people of Neuburxdorf paid little attention to us as we were lined up and readied for our walk to the Stalag.

I don't know whether we first saw or smelt Stalag IVB. I do know that we caught the stench of the place from a long way off. Our first glimpse was of the guard towers spaced evenly round the perimeter. Then we saw it — rows and rows of drab, featureless huts and high barbed-wire fences.

We entered past guardhouses and under a walkway surmounted by a huge metal eagle with a swastika in its talons.

The mere action of passing from the outside world into the confines of the Stalag affected each of us in the same way. There was a noticeable quiet in the ranks — a feeling of unpleasant apprehension as to what lay before us.

Peter Balderston was a wireless air gunner with 466 RAAF Squadron and was shot down over Berlin on the night of 28/29 January 1944 (see Chap. 4). He talks of incidents during his period of captivity and gives a picture of camp life and the battle to fight boredom:

I went, initially, to the interrogation centre at Dulag Luft in Frankfurt. From there, I was sent to Sagan by train. This took about a week in a cattle truck.

It was midwinter and there was snow and beautiful scenery outside. But inside the cattle truck it was murder. I was freezing, half-dead. My trousers had been partly burnt off in the crash and I didn't get another pair for a long time. I was wandering around with one leg out of my pants. Fortunately, I still had my flying boots.

We were only at Sagan for a few days and were then loaded on to another train to Heydekrug. This was near the Baltic, very close to the border with Lithuania. It was very, very cold.

The billets at Heydekrug were four long, low-slung huts. The rooms were twenty feet by fifty feet and held, initially, seventy-eight airmen in three decker bunks. The roofs didn't have a great slope and the snow would lie on them. In the morning, when we got up, the huts were hidden by big mounds of snow and we had to dig our way out.

We weren't allowed to go to the latrines during the night. We had a bucket in the room, but, with over 100 blokes in there towards the end, that got full pretty quickly.

In the morning, one bloke would plough through all the snow, across the parade ground, to the latrines, making a track. Then all the other huts would join into his track. Within half an hour there was a long trench in the snow, with a solid footing.

The latrines were a big ditch in the ground. It would have been 150 feet long with a width and depth of four to six feet. The seats were built over that — two rows of forty. They made a good grandstand for events on the parade ground in good weather.

There were usually two tracks across the parade ground to the toilets. Quite often, in midwinter, all that could be seen of them was the heads walking across. In summer the weather was great. There might be some terrible storms, but apart from that, the weather was beautiful.

Food, its quality and quantity, became an obsession with most prisoners of war. All can recall, vividly, what they had for meals. Peter Balderston is no exception:

Once a week we got about an inch of sausage and each day a portion of bread — one eighth of a loaf, about ten inches long. It was black bread — a dull varnish in colour — mainly made, we used to say, out of wood. One often got a splinter in one's mouth while eating it. The Germans didn't waste flour to roll it with, they used sawdust; and half of it would have a crust of sawdust. It held together better if toasted than if just buttered, so we normally toasted it.

The Red Cross were very good. They provided plenty of butter, jam, tea, coffee, sugar and margarine (such as Alpengate), together with a couple of tins of salmon and a tin of KLIM [milk] in a Red Cross parcel. We used the KLIM tins to make utensils.

We didn't trust the German guards. One said he came from Melbourne — he spoke pure Australian with an Australian accent. He said he'd run an importing business in Melbourne and had come to Germany just before the war started to arrange some imports. But the Germans would not let him return so he stayed and was drafted into the army. We still didn't trust him.

The young ones who had been wounded on the Russian front were the worst. They had been brought up in the Hitler Youth. They were very quick on the trigger.

There were not a lot of people killed in the camp. A couple were killed when they were seen by the guards as they were trying to climb over the wire.

At night, there was often a burst of machine-gun fire followed by a few rifle shots and shouting as someone else had a go.

Various measures were taken to counter the effects of confinement and, particularly, the boredom and frustration. Peter gives some interesting examples:

We had angling societies. There was a trench through our compound. It was three feet wide, at the widest, and had a trickle of water running through it. Ten or fifteen anglers would fish along its length. Then they'd argue about which was the biggest, who'd won, and who'd get the most cigarettes. The biggest fish was about two inches long.

We dug holes in the trench and we had sailing boats and 'putt-putt' boats. The latter were little open boats, in which a candle made from margarine with a wick in it heated air, which passed down a tube as it expanded. The air came out of the back of the boat and pushed it along. These were made out of the milk tins, soldered with the silver paper from the chocolate in the Red Cross parcels.

There were races, with big bets laid. The bets were in cigarettes — a bookie laid the odds. It was called the Royal Motor Boat Club. There was an elected Commodore and President with other officials. The races were held every two weeks or so. They were announced by notices.

There was a chap named Price. He'd talk on any subject. I don't know whether he was well educated, but he was well read. He'd give a lecture on anything. A big room would be crowded and he'd talk from seven o'clock till lights out. Some of his topics were — My Home, 42 years in Africa, Yorkshire Rambling, My Dear Old Ireland, Life in Wales, My Arctic Home, Dust in my Clogs, My Life in South Australia, A Talk of my experiences as a Boundary Rider. I never found out whether he'd been to Australia. I don't think he had, but he could talk about it.

We used to have cricket matches and soccer matches. We played Test matches in cricket — Australia played England. There weren't many Australians — there might have been twenty, so everyone got a game. I've played for Australia in a Test match. We won the Ashes — some charred wood in a paper bag.

There were all sorts of courses from the University of London which could be taken — even farming. There were other courses — English history, poetry etc. Blokes got their degrees — even in farming (with the practical work excluded). Accountancy occupied the time quite well.

There was a Royal Aero Club of model aeroplanes — made out of bed boards. Every walk of life is represented when you get 5000 people in a compound.

One barrack was set aside as a theatre and we put on some beautiful plays. One was called *Grouse in June* — a wonderful play. We all had jobs. I was an electrician backstage.

In the early days, it was just a matter of switching the lights on and off or dimming them a bit. Then I invented a switch that would dim the lights electrically using a series of KLIM tins. They were big one-pound tins.

The switch had a series of these tins with water in them. Above them was an arm with lead weights hanging from wires. Electric current passed through the arm and the wires. If the bar was lowered, the light dimmed as each weight dipped into a KLIM tin. If the bar was raised, the light brightened.

About three one morning, around three weeks after this had gone into operation, there was a great fire and the theatre was burnt down. I don't say it was the fault of my dimmer.

We had quite good talent — lots and lots of people. The most notable name which I remember is Donald Pleasence. He was a WO or a Sergeant.

We had female leads. Wigs were made from wood shavings and the like. Costumes were made from all sorts of things. Our Tailoring Department made amazing things — even German uniforms.

One bloke actually walked out of camp wearing a German officer's uniform made in the camp. He saluted the guard and the salute was returned. He was away for about two weeks.

A craze for hat-making swept the camp. I used the sleeveless pullover I was wearing when I was shot down. A sleeveless pullover wasn't much use, so I pulled it apart, got the wool out of it and crocheted a cap like a Canadian lumberjack's cap, with a peak. I used a filed-down toothbrush to make a crochet hook. I had to learn to crochet. I think I used two-ply wool. I used a bit of tin from a KLIM tin for the peak.

We played ludo and all sorts of games. The Red Cross sent games like ludo, snakes and ladders and cards. We gambled on some of them. We had arguments, even on ludo if we had ten cigarettes on it.

We, particularly the Canadians, skated. The Red Cross sent skates. We flooded the ground and the water froze in a matter of half an hour or so. The Canadians played ice hockey on it.

An unpleasant event was the occasional visit by the Gestapo:
The Gestapo would investigate the camp and have a purge from time to time. I remember an occasion when one of them came on a bicycle. When he went to leave, there was no sign of it.

Later, we found some ingenious person had made a clock out of it. The pendulum went up and over the sprocket.

Escape, or the thought of it, was a boost to morale.
I got out three times. I wasn't in the Great Escape at Sagan, but I did a bit of dirt disposal in my camp. We

Grave of five unknown British airmen at Luckenwalde.

used a pair of old long johns. The tips were cut off and then the bottoms of the legs were tied with bow knots and the legs fed down the legs of your trousers. They were filled with dirt and you would walk around in a crowd of blokes, pull the string to let the dirt fall out, and they would disperse it.

I escaped in the 'shit cart.' It was brought into the camp every day by a Russian prisoner. The cart was a waterproof box. The driver had a dipper on a long stick. This was used as a ladle to transfer the contents of the latrine trough into the cart, which was then driven into the nearby forest and emptied. He did this trip about ten times a day.

I thought that, if I could get under this wagon on the last trip of the day, they surely wouldn't bring it back into the camp. then I would be left outside with it. But they didn't.

I crawled into the box under the container and was there for about two hours. Thinking I was out of the camp, I opened the lid and got out. But I was back in the front lager.

It was dark and I was in terrible strife. The Germans

didn't know how I got there — they had to open the gate to let me back in. I told them I'd been working in the Red Cross store. If you did some work you got extra rations.

I wasn't in the container, but under it and I must have stunk to high heaven. All the chaps in the barracks said they could smell me.

The next time, I got out under the wire. It was winter, and we put a bit of white sheeting over us and crawled and slid our way under the wire.

We got away and were going to head for the Russians — they were coming fairly close. We headed towards them and got under a bridge. There was a lot of fighting going on so we thought we'd better get back.

But then we decided we'd head for the coast, try to pinch a boat and row to Sweden. Just as well we never got to the coast — I doubt we'd have made the sea crossing. About two days later we were walking through a village at about 4 or 5 a.m. It seemed pretty quiet and we were tired, so we decided to walk straight through. We were halfway through when a dog started to bark. The next minute, someone started shooting and, as they were only aiming around us, we put our hands up.

For that I got two weeks in the 'cooler' (the gaol) and then went back into camp.

Peter Balderston showed me a picture which he has kept from his POW days:
You are looking, 'at a picture of the march out of Hydekrug to Iorun in Poland. We went by train. The march is to the station. I can pick out Larry Slattery. He was the first POW. He had been flying Fairey Battles around Heligoland for weeks before the war.

They would often sight German aircraft before war was declared. They would wave at them and waggle their wings. There was no animosity.

They didn't know war had been declared on their last trip. When they waved at the German, he shot them down.

Larry was an LAC air gunner. They didn't have sergeant air gunners then. They just picked them up from the armoury.

The Germans had no POW camp to put him in so he stayed, for quite a long period, in a hotel in Kiel, with a German guard.

Sam Birtles was a warrant officer navigator/bomb-aimer serving on 454 Squadron RAAF when he was shot down near San Marchello in Italy on 28th August 1944. Here is his story:
After I landed, I could hear the Germans coming. I couldn't see them so I took my parachute off, stuffed it under a blackberry bush and started to run. Apparently, they could see me because they started to shoot at me. I decided to be a live coward instead of a dead hero. I sat down and took out my wallet. I had a few hundred thousand lira and was looking to see if I had any photos.

I had photos of my mother and sister and I had two receipts from the squadron. As they had references to the squadron, I tried to chew them up — but I was so nervous that my mouth was dry and it wasn't at all easy.

The Germans arrived and walked me down to the village. They were only boys. I had shrapnel wounds in the left elbow. They started to sprinkle me with powder and bind me up with brown paper bandages. I thought this was pretty inadequate, but then I realised that these soldiers were from the German front line. Then I realised that we were only six miles from it and they were treating me in the best way they could.

I was captured south of Bologna and taken to a big army camp in the hills. It wasn't too bad at first and I used to yarn to the German soldiers. One night, however, a bloke came up behind me, in the shadow, put his gun just behind my head, and said, 'We will blow the bastard's head off.' I don't know if he was joking, but it was a very nasty moment.

Then I was transferred to an Air Force station. They gave me a palliasse in the cell. When I picked it up, it was crawling with bedbugs. I wasn't too taken with that. An Italian guard said, 'Where you from?' I didn't want to say. I thought I was cunning and said, 'Forti,' one of our targets. It just so happened that he'd been there when we bombed it. So, back I went to the palliasse thinking, 'Birtles! you bloody idiot!'

We were only there six days. Then we went up through Ravenna Pass to Innsbruck, and then to Dulag Luft in, where there was a big interrogation centre.

There we were locked in various cells, about seven feet by ten feet. there were marks on the walls where people had marked the days off with pencils. Some had been there for months. It wasn't too inspiring and my wounded left elbow was getting more painful all the time.

I had been there a week when a bloke who could speak English took me for a walk in the sunshine. He asked me, 'Are you interested in radio?' He wanted the call signs and frequencies. I later said to a guard who knew English, 'Your CO must be a dill. I know nothing about radio, but if he thinks that taking me for walks will encourage me to give information, he can forget it. I'm not going to tell you anything.'

I seemed to be shooting my mouth off at the wrong time in Germany. I was called in to the CO who said, 'You won't talk, so you will stay here for the duration of the war.' This was a bit scary.

My left arm was really playing up so I called the First Aid bloke. He was from Chicago and had been called home to help the Fatherland win the war. I got to talk to him and he cut open my elbow without anaesthetics to get out bits of metal. I just fainted — it was so painful.

Anyhow, he must have pulled some strings or something because I finished up in hospital for a few days. Then they took our party by train to where there was an Australian doctor. He took out the shrapnel with anaesthetic and I finished up with my left arm in plaster. He was an army doctor, taken in Greece. There were three of them in the camp.

After about a week, the Germans took us over to Meissen, where POWs. were kept in the Opera House.

I was there for three months. As we were in hospital, we received Red Cross parcels. It snowed at Christmas and we got a sort of cream. We mixed the cream with raisins and jam, put it in the snow to freeze it, and have ice-cream.

I got scabies (crabs) and had to be shaved and dabbed with gentian blue. The fellow in the bunk below me had terrible burns. His hands were all gnarled and his face was horribly burnt. Because of this, he couldn't wash, got the crabs and they crawled up the bedpost to me.

The Allies dropped their troops at Arnhem at this time and there was a flood of people coming in. The hospitals couldn't cope, so they tipped me out. I was supposed to be there for another month — my arm wasn't healed. I had it on a pulley to move it up and down — very basic physiotherapy.

The longer you were in POW camp, the more organised you became. For the last month or so, we were upstairs in the Opera House and had a room to ourselves (two or three of us) working as a combine. One would do the cooking.

We had made machines to provide the heat for cooking. They were blowers with firesticks in the end — you could boil a billy of tea in two or three minutes. The blowers were hand driven fans. By turning the handle furiously, air was forced up a tunnel and past the firesticks at the end.

One way to relieve boredom was to keep a diary. I did this — I used an exercise book and scraps of paper. It was difficult to hide it at times. The guards would kick us out into the open and I'd have to hide it at short notice. They never did find it.

When we went outside the camp — rare occasions — there were dandelions growing.

Some prisoners who were quite seriously injured when they were shot down were repatriated. One young bloke had lost both legs — the First Aid bloke carried him to the transport.

It was sad to see blokes you had been with for months going home, while you stayed. They took with them phone numbers to be rung and letters to be posted.

There were French prisoners near us. They were living in luxury compared with us. I went over there and learnt to play bridge.

While there, I found they had fresh white bread. They were going out to work and meeting people, who offered them bread for the cigarettes they got in their Red Cross parcels.

The people who worked outside could supplement their rations. We were not allowed to work. Only those below the rank of sergeant could do so.

Some of the army blokes worked on farms — some even finished up having a couple of children. They couldn't go home, but they could eat well and have some home comforts. Our only supplement came from the Red Cross parcels and what we could buy with cigarettes.

We were next taken from Meissen to Breslau — right across Germany. It took four days in the train. We camped in waiting rooms — our guards carried sub-machine guns and had dogs. While we didn't like the idea of the guns and dogs, they did, in fact, protect us. We would take over a carriage in a train while the German people were outside, lying on the floor.

The Russians were bombing close to us at Breslau. It got more and more hectic until, after a week, they said, 'Pack up! You're leaving.' This was one of the most amazing things I've ever seen. They told us at about 11 p.m., and by 3 or 4 a.m. we were walking.

We all thought we'd be able to take a case or a bag. But after walking five or ten miles, people started throwing things away. They even threw away things they should have kept — like blankets and balaclavas.

We walked for three weeks. You had to keep up or drop out — there were bodies everywhere. The Germans had a horse-drawn cart with tall wheels. If you couldn't keep going they put you in that. I remember one sick bloke couldn't be fitted into the cart. He probably would have died, but for an English Anglican minister, who carried him all night.

That minister was a help to us all. He encouraged us to keep going with the, quite false, promises of hot soup at the end of another twenty or so kilometres. He was a real hero.

It was hard going through the snow — especially so if you were helping to pull the cart with the sick in it.

We were always hungry. We had four slices of bread each day and some turnip soup at night — although I never saw any meat or vegetables in it. If we got the chance, we'd search among the cattle food in barns which we passed or in which we slept. It was pretty slim pickings.

We'd get to a camp and they'd tell us we could have thirty-six hours rest. But at the end of six hours we'd get up and go on. We didn't know whether we wanted to live or die. It really was the worst feeling I've ever had in my life.

We finally got to a station and there were carriages waiting for us. But no engines could get through because the American Thunderbolts had wrecked the track.

The engines made it finally, and they packed us in. I think there were forty or fifty to a cattle truck. We couldn't lie down. I tried to sit down when possible. There was no water and no toilet facility. We spent three or four days like this, until we reached Luckenwalde — fifty kilometres south of Berlin. We stayed there for the rest of the war.

When we got to Luckenwalde, some prisoners were caught stealing their mates' black bread. The Germans paraded them in front of us. They then had to get them out of the camp or they would have been killed. There was no sympathy for anyone who would steal his mate's food. There were Russian prisoners at Luckenwalde. They did not receive Red Cross parcels. They had to parade while the guards counted heads. To maintain their numbers, and hence their rations, the Russians brought their dead on parade and held them up to be counted. The weather was bitterly cold, so the bodies took some time to disintegrate.

The Germans took three to five hundred of us to Luckenwalde railway station. We were to be taken to Berchtesgarden, to be held as hostages in the peace negotiations.

The Germans made us punch holes in tins of meat which we had brought with us, so that they could not be kept indefinitely. We had to wait at the station until they could bring an engine to take us.

But now, the German attitude was changing. They could see the war was ending. The guards would now carry the sick to the toilet facilities and back. Previously, they would not even provide water or facilities.

We plastered the sides of the train with signs that we were Allied prisoners of war.

When, after a couple of days, they still couldn't get any engines, they took us back to the camp.

There, we slept in one half of a hut. The other half was a store. Someone decided they were going to break in. They tried one night, but didn't succeed. They decided to try again the next night, but the Germans were waiting and, just as the door was broken down, they started firing. Everyone in the hut said, 'Oh, shut up! Let's get some sleep.' I don't think anyone was hurt.

One day, a Russian commander came in and liberated us. It was almost anticlimactic — we'd been expecting it for so long.

We'd been listening to Blood and Guts Patton making forty to sixty kilometres a day, when, suddenly, he

Russians with freed prisoners at Luckenwalde.

stopped. We couldn't understand why. We later found that it was under instructions, because the Russians, English and Americans were going to share Berlin.

The Russians put a guard on the camp and kept us there for a month. The Americans had sent fifty trucks to get us away, but the Russians sent them back, empty. They said they would swap us for the German prisoners which the Allies liberated.

They treated us worse than the Germans had — at least they had given us the Red Cross parcels. You were lucky if a Russian didn't stick a gun in your belly and steal your watch. They arrested one of their number and he had sixteen wrist watches on his arm. I think they were just one above the animal class.

A group of us wandered into a little warehouse and found some Russians who had baled up a Mongolian type who had defected to the Germans. They indicated to us that they were going to shoot him out of hand. We felt this was no place for us. They might not be able to shoot straight. When we had walked about a hundred yards away, they shot him because he was wearing a German uniform.

Out of curiosity, we had a look at the barracks in which the Russian POWs. had been housed. You've never seen such filth and squalor. But, there was a surprise. Although the Russians, as good communists, were supposed to be atheists or agnostics, there was a little room, about twenty feet by ten. It was a Russian Orthodox chapel. It was spotless. There were holy pictures, drawn with crayons.

The Russians insisted that they get the names of all 1500 prisoners of war recorded as held at the camp. Until all were registered, none would be released. Some had already escaped, so we had to double up, to make up the number, and went through the check with any name we chose. Once they got 1500 names they arranged the release.

They finally let the trucks through. The Americans picked us up and took us to Calais. Then they flew us to Brussels, where they let us out on leave. It was the first time we had been free. We kept looking over our shoulders to see if there were Germans about.

We walked down the streets, drinking beer and eating strawberries. The strawberries gave us stomach upsets.

The Yanks could not have been nicer. They fed us white bread, cakes and cutlets — as much as you could eat.

From there, we went across the Channel. It was dramatic to see the white cliffs of Dover It seemed so long since we'd seen them — only nine months actually.

We landed at Horsham, south of London. The NAAFI was there and folks threw there arms around us and kissed us. One old dear put her arms round me and said, 'Welcome home, darling, it's good to have you back.'

We had to be deloused. We had white powder in our hair, under our arms and down the front of our pants. We shed white powder everywhere we went.

I went back to Italy some twenty-nine years later. I went to a pub and an interpreter said, in Italian, 'This man was shot down here on 23 August 1944. Does anyone remember him?'

Italians are very voluble and everyone wanted to get into the act. They all said they remembered. But there were three planes shot down near San Marchello, and their accounts didn't tally. However, one old man came up and said, not only did he remember it happening, but that he had a part of our plane at home.

He took me into his home, and there he had the radio aerial and part of the carburettor of G George. He said he had had a propeller, the starboard one, (the port one was shot up and 180 pieces of it had been in me), but had sold it because he had needed some ready cash.

Dudley Hannaford was a wireless operator in a Lancaster shot down during a raid on a target fifty or sixty kilometres south-west of Dresden on 14 February 1945. His escape from the plane is told in Chapter 5. After he reached the ground:

I saw two other parachutes. I assumed them to be the navigator and another — but I did not dream that the other would be the pilot.

This other chute landed about fifty or a hundred yards from me, and by the time he'd walked to me I had my parachute buried in a ditch. He called out 'Waddington' [the name of their base].

I said, 'Is that you, Brian?' He said, 'No it's Padge.' [the pilot]. I said, 'Oh! You made it?' He said, 'I was at only 500 feet when I jumped.'

The plane had passed us and it was aflame from stem to stern when it exploded, about 300 yards away.

We buried Padge's parachute and started to run in the opposite direction to the plane. I remarked, as we ran, 'I think the fact that I got out means that everybody got out.'

Little did we realise that the bomb aimer and flight engineer were still in the plane when it crashed. This upset us very much when we later found out.

We tried to run unobtrusively but we roused a dog or two. Once one barked, they all did. We tried to avoid going anywhere near a house.

It wasn't long before we came to a fingerboard which said 'Perge 15 km.' We moved off in an ESE direction

German Newspaper Headlines — Invasion, June, 1944.
(23)

Leaflet distributed in prison camp at Belaria - Stalag Luft
III - on occasion of the invasion of Europe, 6 June 1944.

and must have walked ten to fifteen kilometres that night.

It was a dark night with heavy cloud cover but we found a barn. We couldn't see very well but we prised the door open. The pilot was first in and found a hay loft, where we stayed. Neither slept much — when one spoke, the other was awake.

We stayed there the next day. We found we were right alongside a connecting road down to a village, which was not far away. We heard a lot of folk passing by. About nine o'clock that morning, two fellows came in, looked around downstairs, and walked out after a couple of minutes. Padge had been to Germany before the war and understood a little bit of German. He said they were looking for someone — but they didn't think to come and look in the loft.

We tried to plan our strategy. Should we go for the German lines, which were only about thirty miles away? We could hear the guns firing, but we thought the Russians might shoot first and ask questions later.

We thought of Switzerland. We were very close to the Swiss border, but we thought this was too heavily guarded.

We couldn't remember whether our forces had crossed the Rhine in many places, but we thought we should make towards Holland. We figured that, if we could get as far as Kassel, we might be able to get a train going west or northwest towards Holland. So, that was our plan.

When it got dark, the following night, we took off, over the hills, and tried to keep away from towns. We had our maps, a compass, escape rations, a few food tablets (Horlicks), a few water purifiers, a container to drink from and a few other little things.

We walked between fifteen and twenty kilometres that night. When it got to about 3 a.m. we thought we'd better start looking for another barn. We found one in the backyard of a house and were awakened, early in the morning, by the children hitting a ball against the side of the barn.

Not long after that, a fellow came in and straight up the ladder to the left of where we were. Padge was asleep. He used to snore, and was in the middle of one when this happened. I gave him a thump, which woke him up. That was a bit worse, because he coughed, and the fellow came up. We covered ourselves with straw, but he saw us.

We stood up with our hands up in the air calling out 'Kamerad! Kamerad!.' But he said, 'Ruski! Ruski! Sshh!' He was a Russian forced labourer and wanted to know where the Russians were. We were able to tell him they were not far from Berlin. He was pleased to hear that. Padge was a smoker and gave him a couple of cigarettes. We were great friends. He dashed off. We didn't know what he was going to do.

In a quarter of an hour he came back with a couple of pieces of black bread and some terrible stuff in the middle of it. It tasted wonderful.

He didn't come back again, so we pushed off after nightfall.

We had different sorts of experiences. Once, we found ourselves in the grounds of a monastery. We got there by accident. Autumn was coming and the leaves were everywhere. You couldn't walk anywhere without rustling them. We heard someone following us. Every time we moved, the other person moved. Of course, we felt that this was it.

We huddled down and our follower caught up with us. We decided to hide our faces, put our heads down and get behind a tree. To our relief, it was a young boy. He just wandered past us, very slowly, taking a good look at us. We got out of that all right.

On the last night, we were getting rather brazen about the fact that we could wander about Germany and no one seemed to care. Padge had his uniform exposed. I had mine on, but was wearing 'goon skins' (boiler suit issued to RAAF personnel) over it. The idea was that, if I was shot down, I could pose as a gardener, get hold of a rake or a pitchfork, and make my way out.

I took with me as well — a cake of soap, a washer and a toothbrush. My crew used to laugh at me, particularly the navigator. I used to say 'You never know, Don, we

might finish up getting shot down. 'These items were, ultimately, very acceptable. At least I could clean my teeth.

We were walking along the autobahn on the final night before capture, the moon was shining. We thought, 'Ah well, if they catch us on the Autobahn that's just too bad.' As we walked, we could hear the sound of a car in the distance. We thought, 'Who drives a car in Germany to-day?' It would have to be an official car.

So, we ran until we found an overpass. This passed over another road. So we decided to scramble down off the Autobahn.

We settled down under the underpass, where we couldn't be seen. The car stopped right above where we were. In the moonlight, we could see the shadows of two hatted gentlemen talking to one another in German. I didn't understand what they were talking about, but I knew my heart was pounding. We waited and, luckily, they got back in the car and drove off.

Padge was quite convinced they were looking for us. He felt someone might have tipped them off. We had passed a couple of fellows when we were wandering along the Autobahn — they might have reported us.

I had slipped or fractured a disc when I baled out, and my back was causing me a bit of trouble. I was becoming a bit of a burden to Padge because I couldn't quite keep up with him.

Padge suggested, 'What say we give ourselves up? The war will be over soon.' I didn't much like the idea, but with my back the way it was it made sense.

This decided, we took the road to wherever it might take us, to find the nearest town. We were really looking for Fera, and we weren't very far from there. We wandered along the road and came to a township called Ronneburg.

I said to Padge, 'What we ought to do is look for a minister of the church.' We thought we might find some mercy there.

We went into the town square in Ronneburg, it was about ten o'clock at night and there were people moving about the streets. Padge was talking to me in German, and I was replying 'Ja.' and 'Nein.' — very intelligently.

It got to the stage that we felt, 'If you really want us, come and get us.' But we decided not to give ourselves up in the township. We couldn't tell what was a house and what might be a Gestapo HQ

We walked another three or four kilometres to a little village — I don't know the name of it. We saw a church on a corner. It was a timber church and we thought the minister might live next door. We had a look at both sides of the church, but we noticed there was only one house alongside it. There was a sign on the gate, which Padge read. It said, 'Air-Raid Warden.' Padge said, 'Oh, look. Let's give ourselves up here. Let's take the risk.'

So we waited till about ten-thirty, and knocked on the door. An old fellow came out, his cap and his wellingtons on, and well rugged up, it was pretty cold. We gave him the shock of his life.

Padge said, 'Englander — flieger.' It took the old man a little while to contain himself. Then he said, 'Cuminzie.' and showed us inside.

After a little while, the family came down and looked over the exhibits. They sat Padge down in a corner, under a life-sized portrait of Hitler and, to my amazement, Padge looked up, giving the impression he had a cold chill from looking at the portrait.

Straight away, the old chap darted out and came back with a pistol, which he pointed at me. I went white, according to Padge. He said, 'Don't worry.' But I didn't understand it. 'It wasn't me. It was him.' Then Padge said, 'Don't worry. All he wants to know is whether you've got any firearms.' So I had to show him everything I had in my pockets. He did the same to Padge.

After this, our captor phoned the civil police. Padge, being the cheeky fellow that he was, demanded that he have a shave, because we had five or six days' growth. They took us into the kitchen, and Padge started to shave. He had one of the beautiful RAF moustaches, but you couldn't really tell, because of the growth of his beard. He got half of his face nicely shaved when two burly, blue-uniformed policemen came. They would wait for nothing and nobody.

They took us back to Ronneburg. They walked behind us — they didn't handcuff us or anything. As we had given ourselves up, they didn't think we would try to escape. At Ronneburg, we had a cell each.

That night, we had a taste of being bombed by RAF Mosquitoes. It was very close.

They handed us over to the Home Guard next morning, and they handed us over to the Army in the afternoon. We were taken to Jura, which was not all that far away.

We shared a cell in Jura for eight days. We were given no food up to the time we got into the cell at Jura. We still had our escape rations and used these to keep ourselves alive. We lost weight very quickly.

After eight days in the cell, we got to know some of the guards. They were very friendly, and Padge gave one of them his watch. In return, we got better rations — until the sergeant of the guard found out. He was very annoyed.

We don't know what happened to the guard, but the sergeant wouldn't give us the nice things the guard was prepared to give us. We were back on the black bread — and not much of it.

While we were together in the same cell, Padge and I only had one half blanket, which was pretty dirty and blood-stained, and a hard-board surface to lie on. It was bitterly cold, and one of us used to pace the floor while the other lay down and wrapped the half blanket round him in an attempt to keep warm.

One day, Padge said to me, 'I hate the sight of you.' and I said, 'And I hate the sight of you.' I thought, 'I wonder what's going to happen now?.' Our emotions had sort of taken over. Thankfully, we had a sense of humour, and we both burst out laughing. That relieved the tension.

We were taken from Jura to a place called Weimar, which was a Luftwaffe station.

We were given a cell each when we got there. We thought that was great. We had only been there about half an hour when I noticed that there was a machine-gun post just outside my room.

The Americans came over in their Thunderbolts and strafed the aerodrome. There was a bit of a dogfight and I heard a plane come down. Next day, when I was taken to the ablutions, I met the American who had been shot down. But we were not allowed to talk to each other.

After four days, we were taken out, to be transferred to Oberursel, the interrogation centre near Frankfurt. It was a rather interesting trip.

We had two Luftwaffe guards. One was in charge, his name was Frederick, and the other was a little fellow, about five feet high, called Wilhelm. He took very short steps and the butt of his rifle used to drag on the ground, due to his shortness. He could never keep up with us, it was a bit of a joke.

The American was with us, so there were three of us and two guards.

We took a number of days to get from Weimar to Oberursel. It was an eventful trip and we didn't know whether we would make it. There were many rail delays. I recall a train stopped halfway between two stations. It just stopped, for no apparent reason, and showed no sign of moving. We learned, later, that the station we were going to had been bombed. Just as it was decided to return to the station we had left, it was bombed.

We were taken off the train and put on to a utility — a very rare thing in Germany.

We had to carry the guards' haversacks. We took it in turns. As there were two packs, and three of us, there was always one without a pack to carry.

We were herded into a cattle truck one night. It was absolutely packed — standing room only — and we were pushed right up, into a corner. The guard kept saying, 'Kriegagefangenen' (POW) so that people stood back and let us pass until we were right in the corner.

The American was carrying one of the packs. It was pitch black, but I could feel him pulling the pack off his back and putting it on the floor. I heard him undo the straps. Then I heard a little rustle of paper and, a little later, I had a nice piece of German wurst thrust into my hand. He did the same for Padge and himself, then neatly did up the packet and the haversack straps. We had a nice meal.

Next day, we travelled in a different train. It was like a suburban train, with an aisle down the middle. The Germans were sitting on one side of the train and we were on the other. Came lunch time, and the guards opened their bags. Apparently, the food had come from Frederick's bag.

The American, on seeing this, said, 'Look out the window. Don't look over the other side.' We looked intently out of the window while the guard explored the contents of his bag. Then he picked up the knife and saw the remnants of meat on it. The American said, 'Aah. I knew I should have cleaned that knife.' Frederick immediately jumped on the American and the two of us.

We didn't get any more food for the rest of the trip — about five or six days. Up till then we had a bowl of soup each night.

We were given a cell each, when we arrived. My cell was number 51. I can still hear the clink of boots as the Germans marched along.

They called your number, opened the door, and told you to come and be interrogated.

Both Padge and I were at the centre for four days. My cell had soft walls so you could mark the days off with a thumbnail. I noted the records of other occupants. Some were there for twenty days, some longer, and some less. The minimum number seemed to be four.

I had come to a personal faith with Christ early on in the Air Force, and believed in the power of prayer. I thought that, if the minimum time was four days, then I should pray to be out in that time. I really believe that I only spent four days there because my prayer was answered.

I was interrogated by a colonel (equivalent) who spoke beautiful English. He had spent some time in the UK and asked a lot of pertinent questions. He also told me how much he knew about our bombing procedures. As a matter of fact, I think he knew more than I did.

Then he asked me some questions — which I refused to answer. He said, 'I suppose you're interested in your crew?' I said, 'Right.' He then told me of the death of the bomb aimer and the flight engineer.

He said, 'You remember your crew?' I said, 'No.' 'What were the names of your crew?' 'I don't remember.'

He said, 'What is the opposite of today?' I said, 'I don't know.' He said, 'Night. Mr Knight, he was your bomb aimer. You won't see him again.' He produced a cigarette case belonging to Bert Knight. You can imagine how I felt.

Then he said, 'In America, they produce a very good toothpaste — a mouth wash.' I said, 'Oh?' He said, 'Yes. Listerine. Your engineer, Mr Lister, you won't see him any more.' I was just looking into the air above his head. 'Yes.' He produced his identity discs. 'They were in the wreckage of your plane.' All this was very demoralising. Up to then, we had thought they were alive.

Then he wanted to know what I did in civil life. I wouldn't tell him.

On the morning of the fourth day, a very slick civilian gentleman, who spoke perfect English, came in. He was very warm and friendly, but I wasn't too keen on him. After all the pleasantries, he said, 'Oh. Mr Hannaford, what did you do in civilian life? I thought about this, and felt, 'What difference could this make to the war effort?'

So I said, 'I was a clerk.' He then said, 'Good morning Mr. Hannaford. We're thinking of letting you go out this afternoon, with your pilot.' I said, 'Thank you very much.' That was all the information he wanted.

It was bitterly cold in that camp. They only put the heater on for two hours during the day. They took our flying boots and put them outside our door, so we wouldn't escape.

There was a chap next door to me, who had been there for at least thirteen or fourteen days. He was delirious, and talked wildly to himself. They thought I was talking to him, and so, wouldn't let him, or me, have any heat at all.

We now went to a place called Wetzler. It was a transit camp — and a rather terrible place. The fellow in charge was really brutal. It was there, for the first time, that I tasted watery potato soup. That was all we had.

We were there for four days, and then went to Nuremberg.

At Nuremberg, we were put into the last compound to be filled. Initially, it held about 1200 men. We used to get half a horse to eat each day. We would see the horse arrive on the back of a truck. We also got dehydrated potatoes and one or two other things — but not much.

There was no mess as such. All the dixies and boilers were rusted. The place hadn't been used for some time.

Padge, a flight lieutenant, was senior officer. He decided to act. He asked me to help him.

We worked in a sandpit for three or four days, cleaning all the utensils. We finally got them clean and he appointed me in charge of the mess hall.

We had started with 1200 men, but by this time we had 2400, and still got only half a horse per day. There were no Red Cross parcels for some time.

The first thing to come from the Red Cross was boxing gloves and wind instruments. Nobody felt like boxing anyway — we didn't have the energy. Nobody did much with the wind instruments either. But the boxing gloves did come in handy later, when we had to carry things on our back. They were good as pads to stop the cords from cutting your shoulder.

We set up the mess with 2400 people in it. We had insufficient utensils, but, as Red Cross parcels arrived, we cut the tops off tins and used the tins as eating utensils.

All we could do with the rations we were given was dice the meat and make soup. I remember the Americans saying they had hit the jackpot if they got one little dice of meat. No one could possibly have two. A dice was a half-inch cube. The Red Cross parcels were an invaluable supplement.

On one Saturday night, Padge came to me and said, 'Dud. Tomorrow is Sunday. What about having a church service? What about you leading it?' I said, 'No way.' Then I said, 'Let's get the fellows together and see what happens.'

There was one community song book in the compound. At the back were a few hymns. We had non-absorbent toilet paper and a few pencils, which we cut into very small stubs.

We distributed the paper and the stubs to ten or fifteen fellows, and they started writing the hymns on the toilet paper. Thus, there would be a few people who would have the words of some hymns on the Sunday.

On the Sunday morning, a padre walked in. We told him what we had in mind, and he was delighted. So we had a church service — 100 per cent attendance — it was a wonderful event.

Our mid-upper and rear gunners were in the next compound. They were captured before us. Ron was interested in reading the Bible.

He said, 'It was an absolute miracle that I got out of that plane.' I agreed — I couldn't rotate his turret. I had a New Testament with me and he said, 'Could you send it over so we can read it here?' I said, 'Yes, but it's in demand here.'

I didn't know where it was half the time, as fellows would want to read it. He said, 'Could you send it over? We've got a Prayer Book, but I just want to read the Bible.' I arranged this, but we had watch for the guards. There was a warning wire, at least thirty feet from the fence. When the guards weren't looking, I threw it over. I told him he could have it in the morning, but I wanted it back at midday. It was over there in the morning and in our area in the afternoon.

The plane which brought Dudley Hannaford back to Rheims from a German prison camp at the end of the war.

My mid-upper gunner has told me, I didn't see it myself, that, while we were at Nuremberg, he heard the sound of marching feet approaching the camp, and then the call 'Left Right Halt!' — absolute precision marching. He looked over and saw a group of Grenadier Guards. He reckoned they had only one pair of shoes between them.

They had been on a forced march from Upper Silesia. To show the Germans they were not demoralised, they put on a show of marching.

I remember that some of them came into our compound later. They were as thin as rakes. Our fellows got hold of them and gave them food straight away. One fellow could only put one foot after the other in small steps. They gave him too much, and he was as sick as a dog. He was too far gone to take food like that, but I believe he survived.

In the concluding weeks of the war in Europe, General Patton's Third Army was advancing rapidly. According to the Geneva Convention, prisoners had to be kept at least thirty miles from the front. So the Germans put us on the road.

We were sixteen days on the road, marching from Nuremberg to Stalag VIIA (Moosburg). There were 12 000 in the column and our group of 2400 was the last to leave.

On the first afternoon, we had only been travelling for two or three hours, we passed a railway bridge close by. Two Thunderbolts came over, bombed the bridge, flew along our column and decided to strafe us. We all took to the ditches.

An Englishman, Flying Officer Hooke, a great fellow, was very inquisitive. He put his head up, over the edge of the ditch, to see what was going on — and lost it. We had three killed, and others wounded. This was near Feucht. We buried the three alongside the road.

The next day, Nuremberg was bombed, but we had gone a bit further south, and were well out of it.

On the following day, two Thunderbolts came over, but flew across the column and wagged their wings to show they now knew who we were. They came every day after that, but the damage had been done.

Just a couple of nights after that, it rained — absolutely poured. The Germans refused to stop. We went on, right through the night, until 6 a.m. They wanted to carry on then, but Colonel Goode, the Senior British Officer, absolutely refused. They gave us four hours rest and we were off again. We just bivouacked in the mud and slush.

We saw another raid on Nuremberg one night. We had a ringside seat. Thankfully, no stray planes came our way.

There was an American padre with us. He was an old man to us — forty or fifty. As the column moved along, he would literally run to get to the head, then he'd come back slowly, talk to each man, encourage him and see how he was getting on. He did this constantly — how, I do not know.

I went on sick parade one Sunday. I had blisters and blood poisoning. I don't know why I went — there was nothing there to help you, apart from Aspros. I just couldn't walk.

The sick were, that day, put in a bullock cart. It had slanting sides and we sat with our feet dangling over the edge. There were about six of us, being drawn by two bullocks.

It had been raining and the ground was muddy and slippery. It was a dirt road, and we were going up a slight hill. All the fellows were sitting along the side of the road, having a rest. Our bullocks couldn't make it, and the cart just stopped. So we had to get out and push the cart up the hill. None of the boys would give us any help. We received no sympathy because we'd gone on sick parade. What we did get was a lot of ribbing.

As we passed over the crest of the hill, I heard singing in the distance. It was the hymn, 'On a hill far away stood an old rugged cross.' I'd never heard it before. The padre had collected a group together, and was conducting a service.

On Monday 9 April, at 0900, we set off, south towards Neustadt, on the Danube. At 1200, two Mustangs flew low over the column. Fortunately, we had put out a sign 'POW.' At 1230 we had a chow halt. The food issue was, 382 loaves of bread, 12 pounds of Margarine, 2 pounds of salt, and a tiny bit of horse meat (German issue), with an individual issue of a third of a loaf per man per day.

On 10 April, we marched south, towards Pfufferhausen. The column had been split into two parts. One half to go to Obernelsdorf, and the other to Unternelsdorf. By 1600, all the men were billeted in barns in these two villages. No German food was issued on this day. The previous day's issue was supposed to last us.

On 12 April, we remained in these villages. At 1200, an American B-17 passed over and dropped a cluster of incendiaries. They fell just short of Unternelsdorf. Our roll call count was 1794. We had an issue of 200 loaves amongst the men.

There was some distrust when we had to share one Red Cross Parcel between four or five men. Everyone would watch like a hawk to ensure he got his proper share. It was pretty hard to share British parcels. The food really wasn't shareable — it was meant for one, not five.

On 13 April, we marched south again, towards Pfufferhausen. We arrived at Margarethen at 1100 and stayed there overnight. We were billeted in barns. At 1200 we received news of Roosevelt's death — someone had a radio.

At this stage, the guards became very strict. On man was shot in the hip. Bread was down to one-ninth of a loaf per man.

We remained in the village until the sixteenth. Food was a worry, as we were running short of Red Cross parcels. We were allocated only two and a half pounds of potatoes for four or five fellows.

On the sixteenth, we left for Hozhausen. We were now down to one American Red Cross and one French parcel among four men.

On the seventeenth, we marched off to Obermunchen, through very beautiful, but hilly, country. We moved again on the eighteenth, when it snowed all day. We arrived at Riechendorf (Willeradors) and stayed in barns. One British Red Cross parcel between four.

We rested there for a day, then set off, on the twentieth, for Moosburg. On arrival, we had a hot shower, and then billeted in huts. Moosburg was a camp which was built for 40 000 prisoners. By the time we got there, with Russians and others, brought from the Eastern Front, there were 100 000 men in the camp.

The filth and lack hygiene in the ablutions, and the sanitary arrangements, were disgusting. Why typhus and other terrible diseases didn't break out, I don't know. Thankfully, no one had brought them.

Mustang fighter.

It was a great day on Sunday 29 April, when General Patton and the Third Army caught up with us and released us.

There was a battle round the camp for three hours that morning. By 12 noon, it was all over. A tank was blown up at the gate — that's how close the fighting came.

I happened to be the runner to Colonel Goode on the night before. He was in close liaison with the German commandant, and we knew the time of release was getting close. I understand a delegation went into no-man's-land the night before to make arrangements for the camp to be handed over. The commandant wasn't going to put up a fight.

I also understand that the SS had made a demand to take over the camp. They wanted to fight it out.

Moosburg POW camp, where Dudley Hannaford was a prisoner. Prisoners in picture are South African and RAAF.

Apparently the German commandant was a pretty strong man. He refused their request.

The jeeps came in at 1200. Within a very short time, doughnuts and coffee were available from stands, with one or two women manning them. You should have seen the queue. Everyone was on it.

General Patton himself came in about this time. He was absolutely mobbed — particularly by the Americans. All you could see of him was the aerial of his jeep moving along over the top of the crowd. Then, of course, the Germans left us.

The Russians went mad when they arrived, because there were Americans guarding us instead of Germans. They wouldn't let us out. However, they couldn't hold back the Russian prisoners. The Russians were given Moosburg village and the prisoners went in.

The saying for the day was, 'How long is it since we have been captured by the Americans?' We resented the fact that we were still behind barbed wire.

We didn't reach England until 11 May. The Americans took us, after a while, by truck to Straubruge — a Luftwaffe aerodrome. After all the Americans were shifted we were picked up by DC-3s. We spent a few days waiting.

We were flown to Rheims, France (a staging camp), and then to Juvencourt, from where we were flown back to the Old Dart by Lancaster (twenty-five to a plane).

We landed at Tangmere on 11 May. After delousing and medical check, we were taken to the Metropole Hotel at Brighton. We arrived at midnight, but not too late to have a nice big bath.

I think I must have spent about two hours in the bath — where I lay and read my mail — and was very thankful to be back.

Mustang fighter (Imperial War Museum)

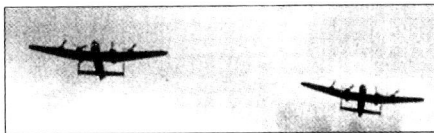

CONTRIBUTORS

(T = Tape   W = Written memoirs   D = Diary entry)

John Appleton
Flying Officer RAF Wireless Operator, Mechanic, AirGunner
578305/56466 210 and 190 Squadrons RAF
Coastal Command   (W)

Peter Balderston
Warrant Officer RAF Wireless Operator/Air Gunner
420516 466 Squadron RAAF
Bomber Command   (T)

Sam Birtles
Warrant Officer RAAF Navigator/Bomb Aimer
418049 454 Squadron RAAF
Middle East/Mediterranean Commands   (T)

Harry Brabin
Warrant Officer RAAF Wireless Operator/Air Gunner
422113 102 Squadron RAF
Bomber Command   (W)

Keith Burns
Flight Lieutenant RAAF Navigator
424361 51 Squadron RAF 466 Squadron RAAF
Bomber Command   (W)

Fred Cassidy
Flight Lieutenant RAAF Wireless/Navigator
35723 30, 93, 457 Squadrons RAAF
South-West Pacific Area   (T)

Geoff Coombes
Flying Officer RAAF Pilot
466 Squadron RAAF
Bomber Command   (T)

Eric Cooper AFC
Wing Commander RAAF Pilot
93 (later 260093) 7, 22, 73 Squadrons RAAF
South-West Pacific Area   (T)

David Corthorn
Flying Officer RAAF Navigator
420452 292 Squadron RAF
Coastal Command and SEAC   (T)

Jack Davenport D.S.O., DFC and Bar, G.M., M.I.D.
Wing Commander RAAF Pilot
403403 455 Squadron RAAF
Bomber and Coastal Commands   (T)

Arthur Doubleday D.S.O., D.F.C., M.I.D.
Wing Commander RAAF Pilot
402945 460 & 467 Squadrons RAAF and 61 Squadron RAF
Bomber Command   (T)

Jim Dransfield
Flight Lieutenant RAAF Pilot
422078 538 Wing RAF, which included 464 Squadron RAAF
2nd Tactical Air Force   (T)

Arthur Chiswell Dutch M.I.D.
Flying Officer RAAF Wireless Operator/Air Gunner
428770 467 Squadron RAAF
Bomber Command   (D)

Ted Eagleton DFC and Bar
Squadron Leader RAAF Pilot
402727 466 Squadron RAAF
Bomber Command   (T)

Andy Emmerson
Flight Lieutenant RAAF Pilot
2239 453 Squadron RAAF
2nd Tactical Air Force   (T)

Noel Gilmour DFC
Flying Officer RAAF Navigator
428664 466 Squadron RAAF
Bomber Command   (T)

Doug Glasson DFC
Flight Lieutenant RAAF Navigator
411695 466 Squadron RAAF
Bomber Command   (T)

Ron Goode
Flying Officer RAAF Pilot
412429 80 Squadron RAF
2nd Tactical Air force   (T)

George Gray D.F.C.
Squadron Leader RAAF Navigator
402346 113 and 55 Squadrons RAF, 454 Squadron RAAF
Middle East/Mediterranean Commands   (T)

Dudley Hannaford
Warrant Officer RAAF Wireless Operator/Air Gunner
432652 467 Squadron RAAF
Bomber Command   (T)

J.R. (Bob) Henderson DSO DFC
Squadron Leader RAAF Pilot
41280 460 Squadron RAAF
Bomber Command   (T)

Gerry Judd D.F.C.
Flying Officer RAAF Pilot
420957 600 Squadron RAF
Middle East/Mediterranean Commands   (T)

Bruce Loane
Flight Lieutenant RAAF Wireless Operator/Air Gunner
421353 466 Squadron RAAF
Bomber Command   (T)

Ron McCathie
Squadron Leader RAAF Navigator
401325 454 Squadron RAAF
Middle East/Mediterranean Commands   (W)

Jim McSharry
Flight Lieutenant RAAF Pilot
404710 217 and 22 Squadrons RAF
Middle East/Mediterranean Commands   (T)

Max McVicar
Flying Officer RAAF Pilot
424818 102 Squadron RAF and 466 Squadron RAAF
Bomber Command   (T)

Peter Matthews
Flying Officer RAAF Navigator/Bomb aimer
424777 454 Squadron RAAF
Middle East/Mediterranean Commands   (T)

John May
Pilot Officer RAAF Pilot
412993 Killed in Action 3/4 December 1943   (D)

Leslie Mills
Leading Aircraftsman RAF Electrician
1542676 1693 Flight RAF
United Kingdom   (W)

Walter Mould
Sergeant RAF
574976 Delivery Flight Malta
Middle East/Mediterranean Commands   (W)

Doug Nicol
Warrant Officer RAAF Navigator/Bomb Aimer
424446 212 and 240 Squadrons RAF
S.E.A.C.   (T)

Bruce Otton
Flying officer RAAF Pilot
424053 466 Squadron RAAF
Bomber Command   (T)

John Page DFC M.I.D.
Flight Lieutenant RAF Pilot
942144 408, 78, 129 and 692 Squadrons RAF
Bomber Command   (T)

Ralph Parsons
Flight Sergeant RAAF Bomb Aimer
409226 466 Squadron RAAF
Bomber Command   (W)

Ross Pearson
Flying Officer RAAF Wireless Operator/Air Gunner
424826 102 Squadron RAF
Bomber Command   (W)

Ted Priest D.F.C.
Flying Officer RAAF Air Gunner
421458 51 and 578 Squadrons RAF 466 and 462 Squadrons RAAF
Bomber Command   (T)

Jack Robertson
Flight Lieutenant RAAF Pilot
421394 14 Squadron RAF
Middle East, Mediterranean & Coastal Commands (T)

Albany (Alby) Silverstone
Warrant Officer RAAF Observer (Navigator/Bomb Aimer)
421405 466 Squadron RAAF
Bomber Command   (W)

Keith Skidmore
Warrant Officer RAAF Air Gunner
421759 78 Squadron RAF
Bomber Command   (W and T)

Keith Spain
Flight Sergeant RAF Air Gunner
3033560 195 Squadron RAF
Bomber Command   (T)

Jack Stronach
Flying Officer RAAF Pilot
402415 117 Transport and 178 Liberator Squadrons RAF
Middle East/Mediterranean Commands   (W and T)

Alan Stutter D.F.C.
Flying Officer RAAF Pilot
421765 463 Squadron
Bomber Command   (W)

John Thomas
Flying Officer RAAF Pilot
424515 102 Squadron RAF
Bomber Command   (W)

Jim Tudberry DFC and Bar
Pilot RAAF
10 Squadron RAF
Bomber Command   (T)

Sam (William) Weller D.S.O., DFC
Wing Commander RAF Pilot
102 and 158 Squadrons RAF
Bomber Command   (T)

Fred Wright, D.F.C.
Flying Officer RAAF Pilot
422996 78 Squadron RAF
Bomber Command   (T)

PHOTOGRAPHIC ACKNOWLEDGEMENTS

The Dennis Hornsey Collection
Bruce Otton
Bruce Loan
Arthur Doubleday
Gerry Judd

Ted Priest
Fred Wright
Harry Brabin
Arthur Dutch
Cyril Johnson

Reg Hackshall
Ted Eagleton
Dudley Hannaford
Syd Kildea
Alysoun Reeves

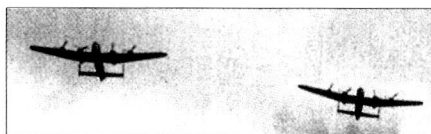

AIRCRAFT

The details of performance, armament and crewing mentioned in this section are taken from the Australian War Memorial publication Air War Against Germany and Italy 1939-1943 and the Memorial's approval to its use is gratefully acknowledged. In most cases the performance is that of the latest mark.

The details of the Demon, Me 210 and Wirraway are an amalgam compiled from a variety of sources, which include crews' diaries etc., and not from any one source.

Anson (Avro)

A twin-engined, low-wing monoplane. Introduced into the RAF as a general reconnaissance/bomber aircraft in 1936, it was obsolescent at the outbreak of war. It was kept in operational use in the UK and Australia for some time, until replacements came off the production lines, when it was widely used as a training aircraft in the Empire Air Training Scheme. It was widely respected for its docile handling.

The Anson had a maximum speed of 178 mph at 7 000 feet and a service ceiling of 18 000 feet It was armed with two .303 machine-guns. Maximum range, depending on load, was 600 miles. Maximum bomb load was about 500 lbs. The crew was usually three or four. (RAF and RAAF)

Baltimore (Martin)

A twin-engined midwing monoplane built as a medium bomber and reconnaissance aircraft for France and Britain, it was only used in the Mediterranean area. It had a maximum speed of 308 mph and a service ceiling of 22 300 feet It could carry 2000 lb of bombs internally. Maximum range was 1100 miles (RAF and RAAF)

Battle (Fairey)

A single-engined low-wing monoplane built before the outbreak of war as a light bomber, the Battle suffered devastating losses in the Battle of France and was immediately withdrawn from operations. It was relegated to use as a training vehicle for use in training gunners and bomb aimers in the Empire Air Training Scheme. It had a maximum speed of 241 mph and a service ceiling of 24 300 feet It was armed with two or three .303 machine-guns. It could carry 1 000 lbs of bombs and had a maximum range of 795 miles. The crew was two or three. (RAF and RAAF training only)

Beaufighter (Bristol)

The Beaufighter had a maximum speed of 345 mph and a service ceiling of 32 600 feet It carried four 20 mm cannon and six .303 machine-guns. Up to 1 000 lbs of bombs or a torpedo could be carried. With normal tankage and a torpedo the range was 1470 miles, this could be extended to 1810 with long range tanks. The crew was two. (RAF and RAAF)

Beaufort (Bristol)

The Beaufort was a twin-engined low-wing monoplane designed for the roles of general reconnaissance, bomber and torpedo bomber. It went into service in October 1939. Its maximum speed varied from 234 to 257 mph It had a service ceiling of 26 000ft. The armament varied but later versions carried nine machine-guns in the nose and two .303 guns in a power-operated dorsal turret. It could carry a torpedo or up to 1610 lbs of bombs. Nominal range was between 1070 and 1660 miles. (RAF and RAAF)

Blenheim, Bisley and Bolingbroke (Bristol)

A twin-engined low-wing monoplane used as a long-range fighter, bomber and in reconnaissance. The Blenheim appeared in five versions during the course of the war. Bisley was the name given to the Mark V Blenheim. Bolingbroke was the name allocated to Mark IV Blenheims produced in Canada. Armament varied from version to version and depended on the task being undertaken, but was similar to that of the Beaufort. In fact, the four aircraft bore a strong family likeness. The Blenheim had a maximum speed of about 270 mph and a service ceiling of around 24 000 feet (RAF)

Condor Fw 200C (Focke-Wulf)

A four-engined low-wing monoplane used as a long range over-sea reconnaissance bomber to shadow and attack Allied convoys. Maximum speed was 240 mph at 14 000 feet Armament comprised two 20 mm cannon and four or six machine-guns. Normal bomb load was 3300 lbs and range was 2200 miles. It carried a crew of eight. (German)

Dakota (C-47A) (Douglas)

A twin-engined low-wing monoplane which was the military version of the extremely successful DC-3 civil airliner. A very reliable and efficient aircraft, it was used in almost every theatre of the war as an unarmed transport aircraft to carry troops and supplies, drop paratroops and tow gliders. The maximum speed was 217 mph at 10 000 feet and the service ceiling 24 000 feet It could carry a useful load of 6700 lbs Crew was three and range was 1600 miles. (US, RAF and RAAF)

Demon (Hawker)

A single-engined biplane developed from a very successful family of prewar light bombers to serve Australia as a fighter and light bomber in the years before the war. It was obsolescent at the outbreak of war but was used at some O.T.U's for pilot training until superseded by the Wirraway. Maximum speed about 180 mph Armament was two .303 machine-guns and one Lewis or Vickers GO gun. A crew of two. (RAAF)

Do-217 (Dornier)

A twin-engined mid-wing monoplane used as a night-fighter, bomber, dive-bomber and torpedo bomber. Maximum speed was 255 mph and the service ceiling was 21 000 feet Armament varied with the role. It was a mixture of 20 mm cannon, 7.9 mm and 13 mm machine-guns. Maximum range was 1440 miles. Crew was three or four. (German)

Flying Fortress (Boeing)

This four-engined bomber was the mainstay of the US bombing effort in the European theatre. It was a very rugged aircraft. In its final form it had a maximum speed of 300 mph and a service ceiling of 36 000 feet It was armed with thirteen .5 in. machine-guns. Its bomb bay could accommodate 6 000 lbs of bombs. It had a range of 2000 miles. The crew was from six to eleven on bombing operations. (US, RAF)

FW-190 (Focke-Wulf)

Single-engined low-wing monoplane fighter used also in the ground attack role. Maximum speed was 435 mph at 39 000 feet Armament was four 7.9 mm machine-guns with an option of two 20 mm cannon. Maximum range was in the vicinity of 1000 miles. Crew was one. (German)

Halifax (Handley Page)

A four-engined bomber, it was also produced in a transport and glider tug version. Maximum speed was about 282 mph and service ceiling about 21 000 feet It was armed with 8 to twelve .303 machine-guns. Maximum bomb-load was 13 000 lbs Range with 6500 lbs of bombs was about 1900 miles. The normal crew was seven. (RAF)

Hampden (Handley Page)

A twin-engined bomber close to the end of its useful career when war broke out. It was also used for minelaying and as a torpedo bomber. Maximum speed was 247 mph and service ceiling was 19 000 feet. It was armed with six .303 machine-guns. It could carry up to 6000 lbs of bombs and had a range of 1800 miles. The crew was four. (RAF)

Harvard (North American)

A single-engined two seat monoplane used as a pilot training aircraft by the RCAF and RAF Maximum speed was 210 mph and it had a range of about 750 miles. It was unarmed. (RCAF RAF)

Hudson (Lockheed)

The Hudson was a two-engined monoplane built as a general purpose aircraft. It was used as a transport aircraft by Coastal Command for general reconnaissance. Maximum speed was 242 mph and the service ceiling was 24 500 feet It was usually armed with seven .303 machine-guns. It could carry bombs or depth charges. Maximum bomb capacity was 1400 lbs. Normal crew was four. (RAF, RAAF, RCAF, RNZAF, US, Netherlands and Chinese Air forces)

Hurricane (Hawker)

A single-engined single-seat fighter, the Hurricane was the RAF's first monoplane fighter. It served in the front line throughout the war. It was the major equipment of Fighter Command during the Battle of Britain. With the steady introduction of the Spitfire and Typhoon in the European theatre, it was used right to the end of the war in the Middle and Far East theatres. The maximum speed 342 mph service ceiling varied up to 41 000 feet Armament was usually eight .303 machine-guns or four 20 mm. cannon. As a fighter bomber it could carry four 500-lb bombs. Its range was about 500 miles but could be extended with external tanks. (RAF)

Ju-88 (Junkers)

A two-engined aircraft used by the Germans as a bomber, dive-bomber, torpedo bomber, night fighter and for reconnaissance. Maximum speed was 370 mph and the service ceiling 30 000 feet It could carry two torpedoes or up to 6400 lbs of bombs. The armament varied with the role - a mixture of 7.9 and 13 mm machine-guns and 20 mm cannon. The range varied between 853 and 1620 miles. A crew of four was used in the bomber, three in the fighter. (German)

Kittyhawk (Curtiss)

A single-engined single-seat monoplane used as a fighter and fighter bomber, the Kittyhawk was one of the American P-40 line which started with the Tomahawk. It had a maximum speed of about 350 mph and a ceiling of 30 000 to 32 500 feet. It carried four to six .5 in machine-guns and could carry up to 1000 lbs of bombs. The range was 470 miles with bombs and up to 1200 miles without bombs but fitted with long range tanks. (RAF, RAAF and US)

Lancaster (Avro)

A four-engined bomber, the Lancaster became the major equipment of Bomber Command. It was contemporary with the Halifax but, although similarly powered, was lighter in weight and had a single large unobstructed bomb bay which gave great flexibility in bombloads. It could accommodate up to 18 000 lbs of bombs and one variant was made to carry the 22 000 lb 'Tallboy' used to sink the Tirpitz and wreck the German submarine pens. Maximum speed was about 287 mph and ceiling about 24 500 feet It was armed with eight .303 machine-guns. Range was about 1660 miles with 14 000 lbs of bombs. A crew of seven. (RAF)

Liberator (Consolidated)

A four-engined bomber used by the US (as the B-24) mainly in the Pacific and Middle East, it was used by the RAF in Coastal Command for reconnaissance, convoy patrol and protection and anti-submarine activity. The aircraft had a very long range when fitted with extra fuel tanks - a useful feature for patrol work and was critical in closing the mid-Atlantic gap at the height of the U-boat menace. Maximum speed was about 300 mph and the ceiling for the Coastal Command versions was about 30 000 feet It was armed with a mixture of 20mm cannon and .5 and .303 inch machine-guns. Maximum bombload was 12 800 lbs. The range of the normal bomber version was 3 000 miles but was greatly extended in the Coastal Command versions. (RAF RAAF US)

Lightning (P-38) (Lockheed)

A twin-engined twin-tailed fighter with a maximum speed of 414 mph and service ceiling of 43 800 feet Normal range was 500 miles but could be extended to 1910 miles. Armament was one 20mm. cannon and four .5 inch machine-guns. (US)

Lysander (Westland)

The Lysander was a single-engined high-wing monoplane built before the war as an army cooperation aircraft. It was built to take off from and land on very short strips and could fly at very slow speed. It had a maximum speed of 224 mph and a ceiling of 23 000 feet It was originally designed to be armed with three .303 machine-guns and to carry 560 lbs of bombs but was probably never used in its original capacity. Range was 590 miles. (RAF)

Marauder (Martin)

A twin-engined monoplane used as a bomber or torpedo bomber. Maximum speed of about 305 mph and ceiling of 32 500 feet Armed with anything from five to twelve .5 inch machine-guns and could carry 5200 lbs of bombs. It had a range of 1800 miles. A crew of six or seven. (US, RAF)

Master (Miles)

A single-engined low-wing monoplane designed and used as a pilot- training aircraft. It had a maximum speed of 226 mph, a service ceiling of 26 800 feet and a range of 450 miles. Crew of two. (RAF)

Me-109 (Messerschmitt)

A single-engined single-seat fighter, it was the German counterpart of the Spitfire and was used throughout the war. The two aircraft were very similar in performance. Maximum speed varied between about 355 mph and 400 mph with a ceiling of 38 500 feet Armament was commonly three 20mm cannon and two 13 mm machine-guns. Range was normally 560 miles. (German)

Me-110 (Messerschmitt)

A twin-engined aircraft originally built as a two seat fighter but could not compete with the single-seaters. Developed as a night fighter and for ground attack. Maximum speed was 360 mph. Armament was a variable mix of 20 mm cannon and 7.9 mm machine-guns. Maximum range was 1300 miles. The crew was two. (German)

Me-210 (Messerschmitt)

A twin-engined low-wing monoplane used as a fighter bomber. It was a development of the Me 110 with improved performance and was itself developed into the Me 410. (German)

Mosquito (De Havilland)

A twin-engined monoplane used as a fighter, fighter bomber, bomber and photo reconnaissance aircraft, the Mosquito had a remarkable performance and versatility. Its main components were constructed from laminated wood. The maximum speed was 417 mph and the ceiling normally 38 000 feet. In the fighter version it was armed with four .303 machine-guns and four 20 mm cannon in the nose. The bomber version was unarmed but could carry up to 4 000 lbs of bombs internally. The range with full bombload was in excess of 1500 miles. The crew was two. (RAF, RAAF)

Mustang (P-51) (North American)

A single-engined low-wing monoplane used as a fighter. In its later versions it proved a valuable escort for the US day-light bombing offensive deep into Germany. It was armed with four 0.5 inch machine-guns and had maximum speed of 486 mph Its service ceiling was 42 500 feet and its range 2160 miles. As a fighter bomber it could carry two 1000 lbs bombs. (US and RAF)

Oxford (Airspeed)

The Oxford was a wooden twin-engined aircraft extensively used as a training aircraft in the Empire Air Training Scheme. It could be fitted for various forms of training. It was not as forgiving as the Anson for pilot training - if it got into a spin it would not come out. Maximum speed (with Armstrong Siddeley Cheetah engines and no turret) was 184 mph and service ceiling was 19 300 feet Range was 874 miles.

Spitfire (Supermarine)

The Spitfire was a single-engined single-seat fighter which formed the backbone of Fighter Command from the end of the Battle of Britain till after the end of the war. It was mainly used in the fighter role but was also used as a fighter bomber. The maximum speeds varied from 355 to 454 mph The ceilings varied from 34 000 to 43 500 feet. Armament was a mixture of 20 mm cannon and .303 in. machine-guns. It could carry a bomb load of 1 000 lbs. (RAF RAAF)

Stirling (Short Brothers)

The first of the four-engined bombers to go into service with the RAF. Maximum speed was 285 mph and service ceiling was 18 000 feet Its performance was quickly surpassed by the Halifax and Lancaster and the Stirling was phased out to the transport and glider tug roles. It was also used as a training aircraft in introducing crews to the operation of four-engined aircraft. Armed with eight .303 in. machine-guns it could carry 14 000 lbs of bombs and had a range of 2500 miles. (RAF and RAAF)

Sunderland (Short)

A four engined flying boat used by Coastal Command for anti submarine patrols, maritime reconnaissance and convoy protection. Maximum speed was 247 mph and the service ceiling was 17 200 feet. Armed with eight to twelve .303 machine guns. It could carry up to 2000 lbs of bombs or depth charges. Maximum range was 2530 miles. crew ten or eleven. (RAF and RAAF)

Thunderbolt (Republic)

A single-engined single-seat fighter used by a number of air forces as a fighter and fighter bomber. It had a maximum speed of 470 mph and a ceiling of 42 000 feet. It was armed with up to eight .5 inch machine-guns and could carry up to 2000 lbs of bombs. It had a range of 1970 miles. (USAF)

Tiger Moth (De Havilland)

A single-engined biplane primary training aircraft which was used very widely for initial pilot training throughout the Empire Air Training Scheme. A very nice little aircraft to fly. Maximum speed 109 mph and service ceiling 14 000 feet It carried no armament and no bombload. (RAF, RAAF, R.N.Z.A.F. and R.C.A.F.)

Valentia (Vickers-Armstrong)

An open cockpit biplane which was virtually obsolete at outbreak of war. It had a maximum speed of 110 mph and a range of 770 miles. Its use was restricted to some training (e.g., wireless operators) and some communications work.

Wackett (Commonwealth Aircraft Corporation)

A single-engined low-wing monoplane designed and built in Australia to provide a training aircraft. It was used mainly for training wireless operators. It had a maximum speed of 100 mph (RAAF)

Walrus (or Seagull) (Supermarine)

A single-engined biplane amphibian flying boat. Used for reconnaissance and air sea rescue. A very rugged machine, it could be catapult launched from ships and landed in the open sea for recovery. Maximum speed was 135 mph and service ceiling was 16 800 feet. It was armed with two .303 machine-guns and could carry 500 lbs of bombs. Crew of two (RAF and RAAF)

Wellington (Vickers-Armstrong)

A two-engined aircraft used as a bomber, torpedo bomber, anti- submarine aircraft and for general reconnaissance and training aircraft at OTUs. The Wellington carried out the first bombing attack of the war by the RAF and served right through to the end of the war. It was built in a number of versions. Maximum speed varied from 235 to 255 mph and ceiling from 18 000 to 19 500 feet (a version with a pressurised cabin could reach 36 700 feet). Armed with six to eight .303 machine-guns it could carry up to 8 000 lbs of bombs or two torpedoes. As an anti-submarine aircraft it was fitted with the Leigh searchlight and carried depth charges. Crew of five or six. (RAF)

Whitley (Armstrong Whitworth)

A two-engined aircraft used initially as a bomber and later as a transport aircraft for paratroops and also as a training aircraft at OTUs. It was used by Coastal Command as an anti-submarine aircraft. Maximum speed 230 mph and ceiling 20 000 feet Armed with six .303 machine-guns and could carry up to 5500 lbs of bombs. Maximum range was 2170 miles. Crew of five or six. (RAF)

Wirraway (Commonwealth Aircraft Corporation)

A single-engined low-wing monoplane built in Australia as an adaptation of the Harvard. It had been intended as a general-purpose aircraft but disastrous meetings with the Japanese Zero led to its immediate withdrawal to training and some army cooperation duties. It had a maximum speed of 300 mph. Basic armament was two 7.7 mm machine-guns or two 20 mm cannon. Range was 1100 miles and service ceiling was 32 000 feet.

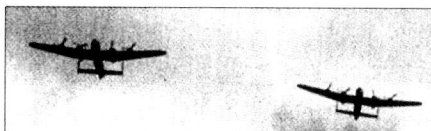

BIBLIOGRAPHY

CACUTT, Len, (Ed.) *Great Aircraft of the World*, Marshall Cavendish, Aldbourne, Wiltshire, 1992.

CHURCHILL, W.S., *The Second World War*, Cassell, London, 1964.

FITZSIMONS, Bernard, (Ed.) *War Planes and Air Battles of World War II*, Ure Smith, Dee Why West, 1973.

HARRIS, Sir Arthur, *Bomber Offensive*, William Collins, London, 1947.

HARRIS, Sir Arthur, Tape Recording of an Address by Sir Arthur Harris to the Members of Squadrons of the RCAF who Served Under Him in Bomber Command.

HASTINGS, Max, *Bomber Command*, Pan Books, London, 1979.

HERRINGTON, John, *Air War Against Germany and Italy, 1939-1943*, Australian War Memorial, Canberra, 1962.

   History of the Second World War: The Strategic Offensive against Germany, 1939-45, H.M.S.O., London, 1961.

   Jane's Fighting Aircraft of World War II, Foreword by Bill Gunston, Military Press, New York, 1989.

LONGMATE, Norman, *The Bombers*, Arrow Books, London, 1983.

MESSENGER, Charles, *Bomber Harris and the Strategic Bombing Offensive, 1939-45*, Arms & Armour Press, Poole, 1984.

MIDDLEBROOK and EVERITT, Chris, *The Bomber Command War Diaries*, Penguin, Harmondsworth, 1990.

MONTGOMERY, Bernard, *The Memoirs of Field Marshall Montgomery*, Collins, London, 1958.

SAUNDERS, Hilary St. George, *The Royal Air Force, 1939 - 45 Volume 3 The Fight is Won*, H.M.S.O., London, 1954.

SAVARD, Dudley, *Bomber Harris*, Cassell, London, 1984.

VERRIER, Anthony, *The Bomber Offensive*, B.T. Batsfor, London, 1968.

WAGNER, Ray, *American Combat Planes*, Macdonald, London, 1960.

I N D E X

Leipzig 67
Liberator aircraft 47, 136
Light Night Striking Force 77-8
Lightning aircraft 79, 136
Loane, Bruce 65-6, 110-12
Low flying 46
Luckenwalde Stalag 111 A POW Camp 121
Ludwigshaven 91
Lysander aircraft 28, 45-6, 136

McCathie, Ron 12-13, 18-19, 37
McSharry, Jim 22
McVicar, Max 40, 76, 97-8, 105
Magdeburg 67, 91
Malta 14-15
Marauder aircraft 42, 136
Master (Miles) aircraft 39, 136
Matthews, Peter 31
May, John 13
Messerschmitt 109 aircraft 53, 78, 80, 96
Messerschmitt 110 aircraft 44, 53, 80, 96
Messerschmitt 210 aircraft 69, 78, 90, 136
Middle East 12
Miessen POW Camp 120
Mills, Leslie 89
Mining 70-2, 97
Moosburg (Stalag V11A) POW Camp 128-9
Mosquito aircraft 68, 77, 79, 106, 125, 136
Mould, Walter 14-15
Mulheim 58
Muhlberg Stalag 1VB POW Camp 109-10, 116
Mustang aircraft 128, 137

Narrandera 22
Navigator 103-4
Navigation School 28-31
Nickels 43-5
Nicol, Doug 31
Nile, River 36
Normandy 73
Nuremberg 68-9, 88

Oberhausen 58
Oberursel POW Interrogation Centre 125
Oboe (target marking device) 57, 79
Observer - see Navigators
Oil targets 94
Operational Fatigue 99-100
Operational Training Units 34-6

Otton, Bruce 32-3, 35, 38-9, 40-2, 91, 94-8, 101-2
Oxford aircraft 25, 32-3, 55

Page, John 47, 56, 71-2, 77-8, 112-16
Paris 42-3
Parsons, Ralph 66, 112-16
Parkes 25-7
Passing-out parades 31
Pathfinder Force 55, 81
Pearson, Ross 11, 25-8, 91-2, 99, 105-6
Persia 18-19
Personnel Reception Centre 32
Pilot training 22-5
Port Pirie 27
'Press on regardless' 88
Priest, Ted 8, 40, 65, 67, 69, 106

Red Cross parcels 109, 112, 116-17, 120, 127-8
Rhine crossing 94
Rhodesia 13
Robertson, Jack 42
Ruhr 51, 56-9, 75

Seagull (Walrus) aircraft 20
Searchlights 51, 76-7, 97
Service Flying Training School 22-4
Silverstone, Alby 9-11, 28-30, 34-5, 42-3, 46-7, 73-4, 98, 100
Skidmore, Keith 58-9, 107-110
South Africa 13
Spain, Keith 97, 106
Spitfire aircraft 39, 137
Stalag Luft 111 (Sagan) POW Camp 111-12
Sterkrade 75, 86
Stettin 51-2, 97
Stirling aircraft 46, 62, 137
Stronach, Jack 16, 46
Stutter, Alan 81-2, 83-5
Stuttgart 78-9
Sunderland aircraft 14, 98, 137

Temora 22
Thomas, John 82-3
Thunderbolt aircraft 121, 125, 128, 137
Tiger Moth aircraft 22, 137
Trondheim 89
Tudberry, Jim 40, 76, 89-90, 97, 100, 102
Tunnels 112
Turkish Air Force 35-6

United States 7-9

Valentia aircraft 27, 137

Wackett aircraft 27
Warnemunde 53
Wagga Wagga 24
Weather 97-8
Weller, William (Sam) 25, 49-54, 56, 59, 63, 76, 79, 81, 92-3, 99-101
Wellington aircraft 16, 37, 42, 44, 49, 53-4, 57, 61, 71, 137
Wesel 94
Whitley aircraft 40, 49, 51, 137
Windfinders 92
Wireless Air Gunner (and Air Gunnery School) 25-7
Wireless Operator 105-6
Wirraway aircraft 23, 33, 137
Wright, Fred 9, 22-3, 75, 80, 94, 96-7, 98
Wuppertal 57

Yale aircraft 24